BUILDING BOTS

BUILDING BOTS

DESIGNING AND BUILDING
WARRIOR ROBOTS

WILLIAM GURSTELLE

CHICAGO
REVIEW
PRESS

Library of Congress Cataloging-in-Publication Data
Is available from the Library of Congress.

For more information, visit www.building-bots.com

Cover and interior design: Lindgren/Fuller Design
Cover photos, clockwise from top left: **Flexy Flyer**, courtesy Mike Konshak and
Team Robot Dojo; **Stinger**, courtesy A. J. Klein Osowski; **Mad Cow**, courtesy
Craig Lovold and Team Rabid Robotics; **Hammerbot**, courtesy Rand Whillock;
Propellerhead, courtesy Mike Konshak and Team Robot Dojo; **Whyatica**,
courtesy Terry Ewert and Team Whyachi. Center photo courtesy **Robot Wars**.
Frontispiece photo: courtesy **Robot Wars**.

Published by Chicago Review Press, Incorporated
814 North Franklin Street
Chicago, Illinois 60610
ISBN 1-55652-459-5
Printed in the United States of America
5 4 3 2

To my sons, Ben and Andy.
They are why I consider myself very lucky.

CONTENTS

ACKNOWLEDGMENTS

Scores of people assisted me in writing this book by providing information and photographs, by answering questions, and by pointing out better ways of doing things. To all who offered help and insight, please accept my thanks.

Robot builders are some of the smartest people in the world. Thanks to the many builders who were kind enough to spend time with me and provide me with their insight and advice. It is impossible to name all of these fine builders who graciously answered my questions and offered assistance, but I thank them all very sincerely.

Special thanks to the people who reviewed sections of the manuscript for accuracy and completeness. Their comments and ideas proved to me that engineers who know their stuff rank at the very top of society's food chain. They are: Ian Burt, Jason Dante Bardis, Dail deVilleneuve, and Stephen Felk. These folks are the heavyweights of the sport.

INTRODUCTION

When I first started to build a fighting robot, it was a frustrating experience, mostly because there were more than a hundred different Internet Web sites with information on how to build a warrior bot. Many of the sites were created by builders who wanted to relate their experiences as they built their own machines. Although parts of the sites were pretty good, trying to put all the information together so I could build a competitive robot was a tough task. It became apparent that new, midlevel, and even experienced builders need a book like this, designed to make building a bot easy and fun.

The popularity of combat robotics has been fueled by television shows such as *Robotica, Robot Wars,* and *Battlebots.* But this sport is not just about what you see on television. In fact, while television made it popular in a hurry, television alone won't keep it going. The sport will grow or wither depending on what happens at the grassroots level. If more and more people build their own bots and stage their own local tournaments, combat robotics could grow into another NASCAR. On the other hand, if it becomes a sport good only for filling up time on cable television channels, the audience will eventually tire of it and it will go the way of the XFL.

Personally, I think the outlook for the sport is promising. Many signs point to a surge in interest at the local level, where it should be. More and more local tournaments are sprouting up, giving builders a chance to compete without traveling cross-country.

So if you want to build a robot, start your own tournament, or just learn more about the sport, you've come to the right place. As far as I can tell, the only absolute requirement for building a fighting robot is interest.

If you check out the pit area—the back room at a warrior robot tournament where builders wire, weld, and fix their robots—you'll see all sorts of different people—young, technical professionals in their late 20s, teams of college students, fathers and sons, mothers and daughters, lone eccentrics, motor sports enthusiasts, retirees from the building trades, and many others. If you enjoy working with your hands and your mind simultaneously, you will find building a competitive warrior robot to be a lot of fun.

Unlike traditional sports such as football, basketball, or hockey, building warrior robots emphasizes intellect over brawn, craftsmanship over coordination, and innovation over physical strength. Here, mental toughness begets metal toughness, and style, creativity, and craftsmanship are as important as winning.

To many, building and competing warrior robots seems like a marriage of motor sports racing and professional wrestling. It is a raucous, young, loud, and action-packed sport requiring nothing but a few dollars (well, maybe more than a few) for equipment and the willingness to experiment and learn new things. I began building robots as a way to spend more time with my sons. Since then, we have both learned new skills, made many new friends, and developed a terrific feeling of pride in our respective abilities to fabricate and build things. Our latest machine is tougher than a prizefighter and more mobile than a gamecock. We're a long way from where we started.

A fighting robot, I think, says a lot about the person who makes it. Some robots are 300-pound metallic boxes on wheels—mean, hulking, indestructible, and free from any pretense to grace and agility. Some are beautiful, catlike creations, able to prowl about an arena like a steel cheetah. Others are hammers on wheels, a platform built to deliver a single deadly blow. All of these robots reflect the style and personality of their creators, so when they fight, their builders are fighting too, delivering the blows and feeling the pain.

That is what makes this sport so much fun. No one gets hurt—

the robots are just hunks of metal. Fighting robots appeal to many different types of people, and that's why I wrote this book: to help more people build their own robots and enjoy the sport of robotic combat.

Who Can Build a Fighting Robot?

You can.

Many people think that building a fighting robot takes an extreme level of technical knowledge and the sport must necessarily be the province of engineering grads and skilled machinists.

Not at all.

All sorts of people, some without any real technical backgrounds, have built good robots, sometimes even great robots. The only requirements are the desire to try new things and the ability to learn.

I wrote this book because I felt that there was no single source of knowledge that a nascent builder could reference in order to get started. To be sure, there are many sources of information available to the builder wannabe. But the information is spread out all over, in too many sources and in too many different places. Often the information in other books and on builder Web sites is too hard to understand and far too dense for the casual or beginning builder to use. Some of it is just plain wrong.

Faced with this, the novice builder's tremendous enthusiasm and excitement soon shrivel. Many new builders who encounter a multitude of motor data graphs, incomprehensible machine design formulas, and page upon page of dense, dry, and prosaic technical information may throw their hands up and decide not to build their bots after all.

Well, no longer, because this book is written for *you*. If you want to get started in the sport and build your first bot, this book will provide you with all the information you need. If you've experimented with robot building but want to reach the next level, this book will help you do so. Even if you have been building for a while and your current robot is made from water-jetted 2021-T6 aluminum, controlled by a self-designed electronic speed controller, and operates a hydraulic lifter arm powered by a four bar linkage, you'll still pick up some new ideas here.

This is a book for all builders. It starts with the basics and breaks down each robotic concept into easy-to-understand chunks. It also explains the technology, in case you're interested, but if you don't care and just want to build your bot, this book will work on that level as well. It tells you how to go about designing your robot, selecting your components, and putting it all together. While the technical aspects can get complicated, they don't have to be. Competitive robots have been assembled in an afternoon from cordless electric drill motors and duct tape, and from modified radio-controlled race cars encased in charcoal grill covers.

Let me share a secret with you: there are only so many ways to build a fighting robot. Sure, all the robots on TV look different and have different types of components, but in actuality they generally follow the same rules and are built in a similar fashion. In this book I explain the way warrior robots are built, from the ground up.

Are you ready to build a fighting robot? Take this quiz and find out:

ROBOT BUILDERS APTITUDE TEST

I know how to drill holes using power tools.

☐ Yes ☐ I can learn ☐ No

I like being part of a team and working with other people.

☐ Yes ☐ No

I'm able to sketch out ideas on paper so other people can visualize them.

☐ Yes ☐ I can learn ☐ No

I know how to solder electrical connections.

☐ Yes ☐ I can learn ☐ No

I can use a joystick to control a video game or a radio-controlled car.

☐ Yes ☐ I can learn ☐ No

I feel comfortable purchasing things over the Internet, from a trusted supplier.

☐ Yes ☐ No

I am willing to spend money on my hobbies.

☐ Yes ☐ No

If you answered either "Yes" or "I can learn" to these questions, you have the potential to build a competitive fighting robot. Answering "No" to any of these doesn't necessarily mean you can't be a builder, but you will likely need to find a partner who can supply the skills you don't have.

No one—not me, not you, not the reigning world champion heavyweight robot builder—knows all there is to know about building fighting robots. Many aspects of this sport change frequently—new equipment becomes available, new methods of doing things come along, tournament rules change, and so forth— so the ability to learn and the willingness to try new things are most important. They're more important than having a machining background or an engineering degree.

How to Use This Book

Chapter 1 is the safety chapter. Because robots can be strong and powerful, you need to make sure you build and compete safely. This chapter is extremely important. Please read it first, and carefully. **Chapter 2** explains the decisions you need to make before you start building. Who will be on your team? What kind of robot should you build? How much money will you need?

Chapter 3 is *very* important! This chapter tells you what is in a fighting robot. The text and diagrams provide the overall arrangement of and relationships among all the components that make up a fighting bot. It explains how a fighting robot is put together.

Chapters 4 through 9 describe the component parts of a robot. As stated earlier, there are only so many ways to build a robot, and almost all robots contain these items: radio control systems, motors, drivetrain parts, motor controllers, and batteries. Each component is explained. While some chapters go into considerable detail, you don't really have to delve deep if you don't want to.

Chapters 10 and 11 describe the materials used in bot frames and weapons. These chapters provide insight on how to choose between, for example, steel and aluminum for a specific part.

Chapter 12 teaches the physics of robot building. If you want to calculate, not guess, how big a motor to buy to reach a certain speed, or if you want to calculate how large a weapon to build to flip an opponent of a particular size, you can use basic physics to find the answers. But if you just want to build your bot, unfettered by theory and algebra, no one will look down upon you if you skip this chapter.

Chapters 13 and 14 reveal the secrets of the different types of weapons that builders put on their robots. Spinners, thwack-bots, wedges, and kinetic energy brutes like spinning disks and cutting blades are described.

Many builders will find **Chapter 15** extremely interesting. It gives builders information on how to incorporate fluid power into their robots. Fluid power refers to pressurized gas and liquid systems that can operate flipper arms, clamps, and hammers.

Chapter 16 provides information, tips, and ideas on how to compete most effectively on tournament day. **Chapter 17** is all about tournaments and competitions: how to compete, and how to put on your own competition. Putting on your own competition is fun and exciting, and easier than you think! **Chapter 18** lists additional resources and where to go for help.

Finally, you'll find a glossary and appendixes that list parts suppliers, a radio frequency chart, and a set of sample tournament rules.

So, if you want to build a robot, you've come to the right place. Ready?

Good. Let's get started.

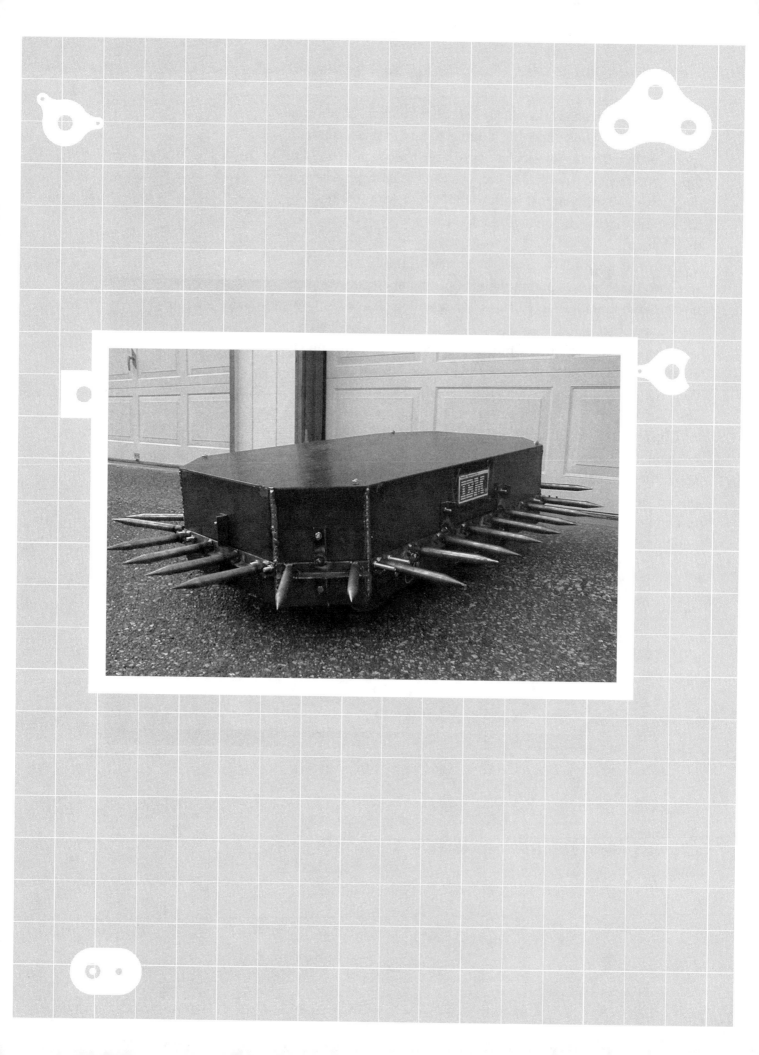

1.
SAFETY

Given the amount of power and the number of dangerous items found on a typical fighting robot, it is almost inevitable that someday, something bad will happen. This chapter provides some basic, commonsense safety guidelines to get the new builder off on the right foot.

The largest fighting robots weigh over 300 pounds and are powered by large electric motors that have enough power to push a Ford truck uphill. These robots are outfitted with power saws spinning at ridiculous speeds, fast-moving hammers, and high-impact kinetic energy weapons. Many robots use high-pressure pneumatic systems that can burst through heavy pipe. Believe me, fighting robots can be as (or more) dangerous to the people in close proximity to them—the robot builders, the robot drivers, and spectators—than to other robots. The other robots are armored with quarter-inch steel plates. People are not.

I have seen high school kids with no training or education in high-pressure pneumatics attempt to compete with extremely suspect liquid CO_2-powered lifting arms. I have yelled at inexperienced builders to not adjust unguarded chain drives wearing long, dangling sleeves. I have heard of close calls when bot drivers attempted to destroy junk, just for fun, in front of spectators, without regard to the inevitable shrapnel produced.

Robot fighting can be very dangerous. If you participate, realize the dangers and heed the instructions

Stinger COURTESY A. J. KLEIN OSOWSKI

given here to mitigate them. Wearing safety glasses and gloves is not nearly enough. No list of safety rules is comprehensive enough to cover every situation.

The suggestions that follow are written to maximize safety. They are only guidelines. Every builder must make safety the number-one priority and accept responsibility to make the bot safe for everyone involved—the builder, the driver, and the spectators.

If you learn only one thing from this book, at least learn this: if you build a robot you do so at your own risk! If you build a robot, it is up to you to build it right and operate it safely. This book will help you. But in the end, all responsibility rests squarely on the shoulders of the builder.

If you are willing to take responsibility for your actions and are still interested in building a fighting robot, then heed these 10 rules. They do not guarantee your safety, but not following them practically guarantees a disaster.

Basic Robot Safety Guidelines

1. **Figure out what you're doing *before* you do it.** If you're going to build robots, don't be half-assed about it—go into it fully assed. Anything less is asking for trouble. If you don't know how to weld, learn how from someone who does. Never used a metal turning lathe? Read up on it. There's a right way and a wrong way to do everything. The wrong way may seem easier, but the right way is better and safer.

2. **Always wear safety glasses.** Almost every shop operation— cutting, welding, sanding, et cetera—involves the presence of flying metal, wood, or plastic chips. When necessary, wear personal protective equipment, such as welding gloves when welding, hearing protection when conducting noisy operations, and a face mask when using paints or solvents.

3. **Get a copy of the rules for any competition you are considering entering.** There is always a section on safety

rules. Read it for two reasons: (1) it provides guidance for building a robot safely, and (2) you won't waste time designing an illegal bot.

Every tournament has its own unique rules regarding safety. The rules pertain to both the construction requirements of the warrior robot and the procedures that must be followed during the tournament. Some tournaments are very particular about the type of radio controls allowed, and some are more concerned about restraints and tie-downs. Because competitions differ, every builder should obtain a copy of the event rules from the tournament organizer well before the actual event in order to make certain the robot is in strict compliance.

4. **Robots are remote-controlled vehicles and they are subject to radio "glitches."** This means they may turn on or off unexpectedly. Glitches occur frequently and come from a variety of sources. Besides a simple signal glitch, your robot may also be subject to radio interference from another contestant's radio transmitter, a commercial radio dispatch system, or even a garage door opener. Never place yourself in a situation where an unexpected radio transmission can activate your robot and hurt you or others.

5. **There must be an easy-to-reach, easy-to-activate main cutoff switch on the outside of your robot.** No matter where you compete, even if you just bang around your garage and scare neighborhood children, your robot must be fitted with a kill switch or removable link. This vitally important safety device is designed such that when it is deactivated or removed, no current flows from the battery to anything else. The drive motors, the weapon motors, the electronics—everything shuts down immediately. This safety device should be completely mechanical; that is, there cannot be any electrical components such as relays or sensors involved.

There are at least a couple of ways of doing this. The first way is to make or purchase a heavy-duty removable link that is easily inserted and extracted from its base. The link is positioned into the circuit in series with the batteries. Removal of the link stops all current flow. The trick is to make it foolproof and easy to pull

1.1 Removable link

out when needed, but not so easy as to allow it to be knocked loose by impact during a match.

The second way is to use a high amperage capacity (sometimes called a high ampacity) cutoff switch. These are available at auto part or marine stores that cater to the racing crowd. These switches often utilize a key or Allen wrench that must be inserted into the switch in order to turn it on or off. Automotive and marine supply stores often sell a variety of battery cutoff devices.

1.2 Cut-off switch

6. **Watch your fingers!** A robot drivetrain may consist of gears, belts, pulleys, shafts, roller chains, sprockets, and what not. These things will remove your finger if they catch it while moving. Be extra careful when working around your drivetrain, and put in guards whenever possible.

7. **Many robots are extremely heavy—use proper lifting techniques.** Be certain your dollies and other carrying devices are rated to accept the weight placed upon them.

8. **Always be attuned to the possibility of shrapnel.** Whenever your robot charges into something, robot parts, steel wreckage, concrete chunks, plastic shards, nuts and bolts, pointy things, sharp things, and heavy things come flying out at random trajectories and at high speed. If it happens in the arena behind Plexiglas, that's probably OK. If it happens in your driveway with the neighbor's kids watching nearby, that's real trouble.

9. **Working with high-pressure fluids requires caution, an understanding of fluid power, and great attention to detail.** Many robot builders use fluid-powered lifting arms or poking devices to make great weapons. Become familiar with fluid power by reading Chapter 15 in this book, and talk to experienced builders or engineers to make sure you are proceeding in a safe and proper manner. There are plenty of other good weapons systems that are less complicated to construct.

10. **Guards should be placed securely over all hazardous surfaces until the robot is in the battle area.** Many builders affix sharp, pointed weapons to their mechanical warriors. The guards should be fabricated from study plastic or other suitable substance and held tightly using a bungee cord or other securing device.

2.
HOW TO DESIGN A ROBOT

Brainstorming

The first thing to understand about robot building is that it is all about trade-offs. This is what makes robot building so interesting. While a big budget is definitely a major advantage, there is no way to build *the* single, ultimate fighting robot, because there's no getting around certain trade-offs.

Trade-offs are the decisions you have to make that force you to "go weak" in one area when you "go strong" in another. No single robot can be strong in all areas. Robot building is essentially an exercise in optimization and choosing between mutually exclusive alternatives. Here is a listing of some of the trade-offs:

- Speed trade-offs
    Speed versus torque
    Speed versus driving control
- Time trade-offs
    Time to design versus complexity of weapon and drive systems
    Time to build versus complexity of weapons and drive systems
- Weight trade-offs
    Motor power versus weight
    Armor thickness versus weight
    Weapon power versus weight

A1 COURTESY TEAM BOBBING FOR FRENCH FRIES

- Cost trade-offs
    - Cost versus weight
    - Cost versus component quality
- Energy storage trade-offs
    - Capacity versus weight
    - Capacity versus recharge time

A robot that is faster, more maneuverable, lower in cost, more heavily armored, and can push harder than any other robot at the tournament is perfect—but impossible to design. A crap-bot that is slow, weak, and cheap is easily designed and built, but what's the point?

The unalterable truth is that no one robot can have it all. So the first step is to decide what you want your robot to do. Many builders are inspired by robot designs they see on television. They sketch out what they like about a robot and perhaps decide how they can improve upon it. When someone comes up with a great new idea, such as gyroscopic based steering control, or aluminum alloy armor, other robot builders jump in and copy the idea. This is OK, as it moves the *entire sport* upward.

Robot Fighting Styles

Besides the physical trade-offs of size, weight, power, and speed, there are style trade-offs as well. Early on, the smart builder will make a reasoned, conscious decision about what sort of fighting style the robot will assume, and then optimize the physical attributes that maximize that style's effectiveness. An example or two will clarify this point.

Two robots face each other at the start of the match. When the starting flag goes up, the robots accelerate as fast as they possibly can, plunging head on into each other. Parts fly off, belts rip and shred, wheels spin unattached across the arena. The stronger bot emerges victorious and the weaker bot is left immobile in the ring. These matches don't last long, but the crowd usually loves them. I call this style of fighting "robo-jousting." This is a good strategy for fast robots with at least some forward-facing armor, small battery packs, and/or large kinetic energy weapons.

In another type of match, the robots circle back and forth, across the arena. They jab and parry but don't engage right away. At some point they lock horns and begin to push. They push with everything they've got, going right up to motor stall torque. Eventually, one robot's motors overcome the other's and it pushes the opponent into a kill saw or a fire-belching pit. I term this "sumo-style" fighting. The greatest sumo wrestler in the world is Hawaiian-born Konishiki, who is a giant—big, hulking, and exceedingly strong. A sumo-style approach is a good choice for Konishiki-like bots: heavy and powerful with real torquey motors and big batteries.

Of course, there are many different styles of fighting, and each style requires a different robot design and different components. Decide what you want your robot to be and design it based on that decision.

The Robot Design Process Explained

Imagine that you're sitting with friends at the local coffee shop, in your school cafeteria, or at the neighborhood pub and you're discussing what would make a kick-ass, hell-of-a-cool fighting bot. One person volunteers to take notes. The next day, a sketch is circulated:

As you can see from this sketch, you estimate that this robot design will:

- Weigh about 225 pounds based on six driving motors, drive control, armor, weapons, wheels, and frame.
- Have a high-pressure, air-powered lifting arm that can toss a 150-pound opponent nine feet up into the air.
- Have a maximum speed of 16 MPH.
- Be able to deliver 650 inch-pounds of torque at the wheels.
- Have dazzling strobe lights to disorient other drivers.
- Protect itself with ¼-inch titanium skirting on all sides.

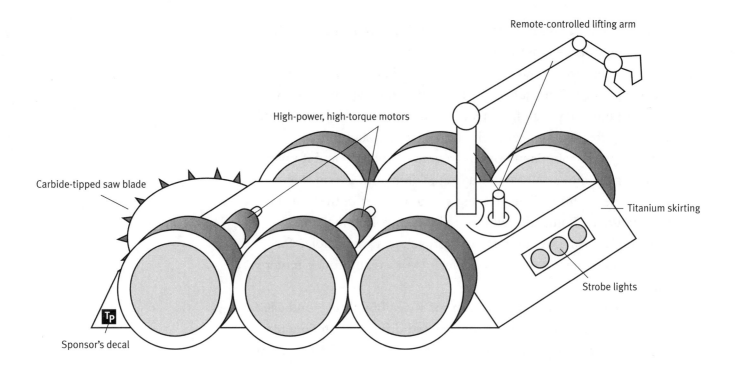

Remote-controlled lifting arm

High-power, high-torque motors

Carbide-tipped saw blade

Titanium skirting

Strobe lights

Tp

Sponsor's decal

2.1 An unlikely first robot

◯ Have a movable, cranelike arm capable of picking up an opposing bot and dropping it on arena hazards.

◯ Use a 10-inch carbide-tipped saw blade for ripping through the opposing bot's armor.

This robot is too expensive for you to afford, but you will get sponsors to pay for most of it.

This would be a great robot all right, but there are a few big problems:

◯ The weight estimate puts us near the bottom of the heavyweight weight class. This means you will go up against robots that weigh as much as 100 pounds more!

◯ Further investigation into the lifting arm shows that no one in the group has any experience with high-pressure pneumatics.

◯ Strobe lights are prohibited by tournament rules.

◯ The titanium costs a *lot* of money and is hard to machine.

◯ The movable cranelike arm requires multiple precision-made four bar linkages and exceedingly complex controls. This will blow the budget and take more time to build than you can afford.

- The carbide-tipped saw blade engine is too big to be contained in the robot body.
- Nobody will sponsor and give money to a bunch of untested robot builders. So you go into round two of the design discussion and keep what you like, throw out what won't work, and modify the rest.

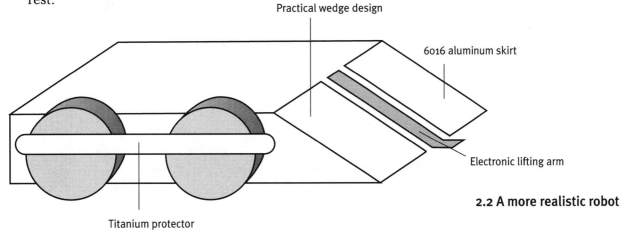

Practical wedge design

6016 aluminum skirt

Electronic lifting arm

2.2 A more realistic robot

Titanium protector

- You cut it down to four motors and wheels and go down in weight to qualify for the middleweight bracket.
- You only use titanium on the sides of the bot, and minimize the machining required.
- You trash the strobe light, movable crane, and carbide saw ideas.
- You decide that a lifting arm is good, but settle on a more practical design for it.
- You decide on a budget for this robot of about $1,600, which you can afford by yourself.

Setting Up a Team

Early on, you need to assemble your team. How big a team do you need? It all depends. Excellent robots have been made by individuals, two partners, trios, and groups approaching the size of Mike Tyson's entourage. The right number depends on what the individuals know and what each can contribute. Usually, each person wears a multitude of hats. In general, teams, whether they consist of one member or eight, need to fill the following roles:

A WORD ON PERSONALITY

Your robot needs a personality. In order to impress judges, spectators, your friends, and the other bot builders, your robot must be more than just a metal, rubber, and Lexan box. Make your fighting robot say something subtly or boldly about you and your team. Make it all-American with a red, white, and blue paint job and have it fly the flag when you enter the arena. Make it black and yellow and vaguely resembling a bumblebee. Make it something extraordinary and cool. But don't make it just another gray, steel, wedge-shaped bot with big motors, big torque, and nothing else. Building a fighting robot is your chance for radical self-expression. Win the hearts of the audience. Make people remember it— win, lose, or draw.

Head Designer

The head designer has overall responsibility for the electrical and mechanical engineering of the entire robot. This person should be familiar, or willing to become familiar, with DC motors, radio control systems, electronic speed controllers, mechanical driveline systems, and basic mechanical engineering principles.

Head Builder

The head builder is responsible for taking the design from the design team and turning it into reality. This person should know about, or be willing to learn about, metalworking, welding, the material properties of metals and plastics, and basic shop practices.

Scrounger

It's not always easy to find the parts you need. That's why a good scrounger is worth his or her weight in gold (or robot parts). The scrounger may be able to find the high-power DC motors that your team needs on the Internet. The scrounger may know someone at a scrap yard who deals in high-strength steel alloys. The scrounger understands materials, knows what works and what doesn't, and realizes what things are worth.

Driver

It looks easy to drive a robot, but it's not. During combat conditions, it takes a cool head, a firm hand, and a steady eye. Anyone can drive a car, but it takes a Jeff Gordon to consistently win on the Winston Cup circuit. It takes practice and commitment to be a really good driver. Find one or become one if you want to win.

Artist

Your robot is the ultimate vehicle for self-expression through technology. Make your bot look so cool that people will chant its name like fanatics at a boxing match. It should express what your team is about through its colors, design, and form. Having an artist or industrial designer on your team can give your bot the look you want.

Assistants

Besides the key people above, there is ample opportunity for others to get involved as assistants, grips, pit crew, general hangers-on, and posse members.

Working Up the Design

Once the team and the general idea for the bot is agreed upon, what next? It's time to do a critical evaluation of the robot concept. A robot evaluation involves a long, hard look at what you're about to do. Every team has to evaluate what's most important to them — remember those trade-offs talked about earlier. In general, the evaluators need to answer these questions, honestly and correctly.

2.3 Titanium monkey robot with prehensile tail

1. **Is the robot designable?** Yes, a titanium robot that hops around the arena like a mechanical rhesus monkey and is outfitted with a prehensile flail tail *would* be a pretty cool robot. But who on your team can really design one? If this is your first robot, you might want to go for a simple and straightforward design.

2. **Does your team really have the skills on to build this robot?** If your design calls for welding exotic metals like titanium or aircraft aluminum alloys, you might want to think twice unless someone on the team has experience in doing this. Maybe you can design your own electronic speed controller out of surplus integrated circuits, but can you guarantee it will stand up during battle conditions? Only do what you feel certain your team can do reliably.

3. **Will it be simple to operate and drive?** Make it a no-brainer to operate in the arena. When fighting, you want to do as little thinking as possible. Don't design in anything that you have to worry about during a match. For example, you might have a lifter that becomes vulnerable when activated to the up position. It might open a hole in the body that provides access to the guts of your bot, and that's not a good idea.

4. **Can you afford it?** A heavyweight robot can easily cost $3,000 without labor. If you add better motors, thicker armor, and more reliable electronics it will go far beyond that. There are ways to economize, but you need to think about that early on. The costs can range from about $680 to nearly $10,000 depending on the size and complexity of the robot. Exotic weaponry such as pneumatic lifters or kinetic energy weapons add even more.

5. **Will it be dependable?** Nothing is more important than dependability—not torque, not speed, not maneuverability, not weaponry. Many, if not most, matches are decided by whose robot can stand up to hits and whacks best. Design your bot first and foremost to be dependable in the battle arena.

6. **Can it win?** Visualize how your robot will fight against a variety of other robots. Determine what your robot has that gives it a chance to win, and maximize that trait.

7. **Most important of all—is it cool?** Remember, this robot is an anthropomorphized metal extension of you and your teammates. Will people remember your bot, even if it doesn't win? Pleasing the crowd with a hip, bitchin', cool-looking robot is as important as winning. Take a tip from pro wrestlers—it doesn't have to be nice, it doesn't have to be slick, it doesn't have to be pretty. It has to be *cool*.

A Look at Costs

Robot building, especially in the mid- to heavyweight classes, is not an inexpensive undertaking. Here is a rudimentary budget for three different sizes of robots without a kinetic energy or lifting weapon. Because these are simple bots, these numbers are modest.

A top-flight combat robot can cost well over $10,000 when the fabrication costs are included. If these numbers are beyond your budget, don't despair. Many terrific lightweight and very lightweight robots have been built on small budgets. The costs go down dramatically for smaller robots. Plus, a good scrounger may be able to find used parts at a fraction of these costs.

LOW-COST, LIGHTWEIGHT ROBOT BUDGET

Parts	Costs
(1) FM used 4-channel radio control transmitter and receiver	$150
(2) DC electric drill drive motors	$150
On-off control relays	$70
Surplus driveline parts including mounting for drill motors, tires, shafts, and fasteners	$100
Sealed lead acid batteries	$60
Surplus steel frame	$50
Miscellaneous electronics	$50
Miscellaneous mechanicals	$50
Total Budget	**$680**

TYPICAL MID-SIZE ROBOT BUDGET

Parts	Costs
(1) Radio control transmitter and receiver	$300–$500
(2) DC drive motors	$350–$700
Electronic speed controller capable of handling two motors	$ 400
Driveline parts including bearings, belts, pulleys, and shafts	$150–$300
NiCad or low-resistance, sealed lead acid (SLA) batteries	$100–$400
Steel and/or aluminum frames	$200–$300
Miscellaneous electronics	$100
Miscellaneous mechanicals	$200
Total Budget	**$1,800–$2,900**

HIGH PERFORMANCE, HEAVYWEIGHT ROBOT BUDGET

Parts	Costs
(1) 8-channel PCM R/C transmitter and receiver	$550
(4) DC 4-horsepower gear motors	$1,500
High capacity electronic speed controls	$2,200
Driveline parts including wheels and axles	$1,000
NiCad batteries	$1,000
Chrome-moly steel, titanium, and/or Lexan frame	$1,000
Miscellaneous electronics	$250
Miscellaneous mechanicals	$250
Total Budget	**$7,750**

Building on the Cheap

If you go the low-cost, scrounger route, then the availability of parts will determine the final design, rather than vice versa. When cost is no object, the builder starts off with a design and then goes out and acquires the parts, paying whatever the market determines. On the other hand, the scrounger bot builder will find some parts and then figure out what to do with them.

For example, if you come across some really good motors, really cheap, then the low-cost robot builder designs around them and tries to find other low-cost parts that will be compatible. The designer must be resourceful and flexible to figure out how to incorporate a heap of cheap but useful parts into a fully functioning robot. Ricky Ticki, Bob, and Sparafucile (see photo insert) are examples of decent low-cost bots with some winning records.

Building a Very Small Robot

Some builders enjoy building and competing with very small robots weighing considerably less than the robots seen on television. There are many competitions featuring robots that weigh as much as 50 and as little as a single pound. Building a very small robot costs only a fraction of what it costs to build a 50-pound lightweight robot.

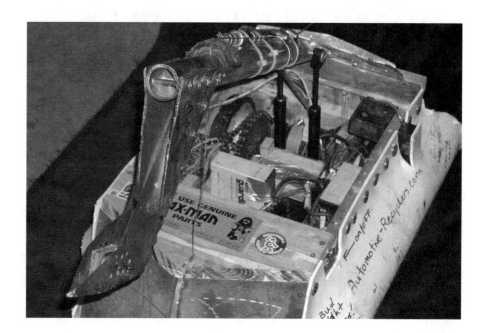

2.4 Bob, a $300 lightweight

Very lightweight robot events are organized by a promoter who decides on the contest rules and weight restrictions and then gives the resulting weight class a name—ultralights, antweights, beetleweights, featherweights, and so on. You can often find information on the Internet for a tournament in your area. If there is no very lightweight event in your area yet, consider starting one (see Chapter 17).

The tournament organizers set forth rules that describe competitions that feature matches between robots weighing under a pound, between 1 and 12 pounds, and robots weighing from 12 to 30 pounds.

For the most part, you can apply instructions and ideas presented in this book to build very small robots. But there are some differences:

1. Very lightweight robots often use 27 MHz radio systems. As you'll learn from Chapter 4 describing radios, the 27 MHz systems are the AM radios that come with toy radio-controlled (R/C) cars. While most builders of larger robots can't use these cheapies, they often work just fine for the very small robots. These radio systems are very inexpensive when bought new at toy stores and are close to free if you can find one at a garage sale.

2. Very lightweight robots often use a modified R/C servo motor to drive the robot. Servo motors, as described in Chapter 5 on motors, are built to position a throttle or other control device precisely. So you'll need to modify the inexpensive R/C servo in order to make it run continuously. This is not hard, but it requires removing parts from the servo and rewiring the inside. Exact instructions depend on the brand of servo you have. (Information on rewiring servo motors is usually available from your local R/C supply store or can be found through the Internet. You should search on "modifying servo motors.")

3. The speed controllers on very lightweight robots cost much less than the ones that control heavyweight robots. For example, the modified R/C servo motor typically used to drive an antweight is controlled directly by the output signal from the radio receiver. There is no electronic speed control (ESC) to worry about at all— just plug the servo into the receiver and you're ready to go.

4. Building a very small robot can be an easy way to get started in this sport. Some builders get their first introduction to combat robotics by simply modifying an R/C toy car by gluing on armor and weapons. It doesn't get much cheaper or easier than this!

Basic Robot Design Guidelines

Following these rules will help you attain robot building excellence.

1. **Sketch it out.** Put your ideas down on paper prior to building. Many issues and problems become apparent during the sketching process.

2. **Make a parts list.** The list should include manufacturer information, cost, performance specifications, and weight— *especially* weight. The weight limits are whole and inviolate. They are complete. They are total. Many a robot builder has built a robot that would have been really, really competitive if it were in the weight class the builder had originally intended. Unfortunately it was three pounds overweight and got squished in 23 seconds by a much bigger robot in the higher weight class.

3. **Consider making a three-dimensional model,** especially if your design is unique or complex. You can make it from foam or cardboard. It's often hard to visualize the interaction and placement of parts from a sketch or even an engineering drawing.

4. **Buy the best motors you can afford.** Good motors are the foundation upon which all else rests. People can win with hokey electronics, or even inoperative weapons. Good motors make good robots.

5. **Design a safe robot.** Make your robot safe for spectators, team members, and all humans and living creatures. Use cutoff switches and high-quality radio systems, and secure your joints and connections. File burrs off, and remove jagged edges.

6. **Learn the skills you need to be a self-sufficient robot builder.** Learn to work metal correctly, figure out how to weld, and understand basic electronics. Not only will it result in a better robot, it will make you a more interesting person and a better guest at dinner parties (at least for other robot builders).

7. **Nothing (except for safety) is as important as dependability.** The match doesn't end early if your bot is still moving. Great robots have been defeated by lesser machines because of loose wires, broken connectors, or lost wheels. Make quality a priority.

8. **Learn from each match.** The battle-tested veterans have better bots because they improve their robot after every match. They learn what needs to be reinforced or redesigned. Ask any expert in quality improvement, even the best bots can always get better.

9. **Get the best parts you can afford.** The differences between alloy steel and pot metal become apparent in a hurry during a match. In robot building, there is a high correlation between an item's quality and its cost. For example, expensive industrial batteries don't cost more than inexpensive batteries because of their massive advertising campaigns and celebrity endorsements. They cost more because they are engineered better, so they last longer, weigh less, and have less internal resistance.

10. **Don't cheapen your bot by resorting to duct tape, glue, and so forth.** Screw things down, weld them, or bolt them together. Too much duct tape makes your robot look cheesy.

BOTBASH 2001

October 6, 2001
Phoenix, Arizona

In the razor wire district of west central Phoenix is a broken-down building, called simply autoMATIC. It is an abandoned equipment warehouse in a neighborhood full of other abandoned warehouses, train yards, and flophouses. It ain't much, but on this day, autoMATIC becomes a bustling enclave of young engineers, and they transform it into a society that resembles an amalgamation of scenes from *Blade Runner, Fight Club,* and *Road Warrior.*

Apparently, there are plans to turn autoMATIC into an artist's studio, but for now, it is just a dump. Just getting there is a scary experience. Inside it, event organizer Bob Pitzer has erected a large polycarbonate and expanded-metal robot fighting area. The mood he's going for is definitely post-apocalyptic cockfight, and he's successful in attaining it. The youngish crowd of 150 or so metal-bending robot builders surround the fighting pit, where big-name fighting robots such as Ziggo, Toe-Crusher, and Backlash fight to occasional mutual destruction.

The builders are a little different than the people normally found on the streets of Phoenix—they are the would-be geek types with an edge about them. They're mostly T-shirt clad, and many of the shirts sport the names of their teams and robots. During the matches they surround the pit on all sides, intensely watching the metal-against-metal carnage within. After a really, really good match, they bang loudly and rhythmically on the metal fence encasing the pit, in praise of the good fight. The feeling of excitement down near the fence is palpable. It's a new type of entertainment, very visceral, kind of like watching a live version of *Mad Max Beyond Thunderdome* that doesn't suck.

Next to the fighting pit are the work areas—long tables lit by portable halogen work lamps whose harsh glare fills the warehouse with giant moving shadows. The builders hunch over their creations borrowing hammers, arc welders, and soldering guns from one another in order to put their freshly wrecked robots back into fighting shape before the next match. The builders are resourceful and can often take a bent and broken robot, weld a patch on, repair the wiring, and it's good to go.

On this night, Ziggo, Locust, and Delta Raptor are the bots doing the best. Ziggo's spinning steel disk makes minced metal out of early round opponents.

He meets his equal in Toe Crusher, a battle-hardened thwack-bot from a veteran builder who was formerly a Walt Disney special effects expert.

Locust is a steel rectangle on wheels. There is no discernable weapon on Locust, like a pneumatic lifting arm or a spring-operated hammer. What Locust does have is raw power, and it's got lots of that. Locust's DC drive motors appear to be huge for a 60-pound robot, and they are probably putting out close to 400 foot-pounds of torque at stall. Locust pushes harder than a copulating elephant. It rams into its opponents with an immense quantity of momentum, and irrespective of their weapons, it simply squashes them into the pit walls.

As the day wears on, the bots become more and more mangled. Late into the evening, at the end, it's a battle of attrition. The long, pointy parts get broken off, the chains that turn the carbide-tipped saws are destroyed, and the CO_2 pressure systems that operate the lifting weapons are broken beyond repair. The hardest bots, like Pitzer's Delta Raptor, are made from exotic materials such as aircraft aluminum and titanium alloys. These are the ones most capable of taking and returning punishment. At the end of the day, these bots take the top prizes.

After midnight, BotBash becomes a different kind of event. A show by underground performance artist Christian Ristow is staged in the adjacent work yard. He and his Los Angeles–based stage crew are clothed in studded black leather, and their skin is machine-decorated with tattoos. Ristow is an alumnus of Survival Research Laboratories, famous, at least in some circles, for middle-of-the-night, ad hoc, machine-art performances. Like a rave for technoids, SLA events are loosely organized, word-of-mouth affairs presented under bridges and viaducts in San Francisco when the cops aren't looking. Today, Ristow and crew have spent the day assembling a collection of steel, animal-shaped robots of immense power. Built from heavily modified earthmoving equipment like backhoes and front-end loaders, the creeping, crawling, walking steel carnivores are scheduled to robotically demolish 10-foot-high pyramids of television sets, stacks of light fixtures, a collection of boxing mannequins, and 20-foot-high plywood cutouts of Jerry Falwell, Saddam Hussein, George Bush, and Osama Bin Laden.

The performance starts late, about 30 minutes after midnight. A group of young, black-clad, heavily pierced 20-year-olds pay $10 to join the robot builders in watching the spectacle. On a raised platform are three bare-breasted young women with angel wings taped to their shoulders. They may or may not be part of the performance—it's hard to tell. The robots roar to life, radio controlled by the shadowy stage crew manipulating radio transmitter joysticks off in the shadows of the yard. Their arms and teeth rip, saw, and chew the industrial scrap and symbolic "art" placed there for them to wreck. The destruction goes on and on, accompanied by industrial-techno-metal music blaring over a large bank of speakers. The performance takes about an hour. At the end, one of robots, a tracked bulldozer with a three-pronged hydraulic pincher called "The Subjugator" finishes off the pile of broken, machine-age effigies with a very long, sustained blast of fire from its built-in flamethrower. The whole audience can feel the hotness of the blast, even through the heat of the Phoenix night.

The 20-year-olds applaud and shout. The robot builders mostly just stand there and look, first at each other, then at the angel-winged girls, then back at each other. They seem unsure as to what just went on. As far as they can tell, the message appears to be something on the order of "robots good, politicians bad," but there may have been a deeper meaning that the pragmatic engineers just didn't get.

3.
THE PARTS OF A FIGHTING ROBOT

The Basic Building Blocks

The credit for thinking up the idea of building fighting robots is often given to Marc Thorpe, a California toy designer. He was experimenting in his kitchen with a radio-controlled (R/C) vacuum cleaner when it occurred to him that an R/C fighting robot would be much more fun. He showed his friends and they began building robotic warriors. People came to hear of Marc and his robots by word of mouth and the sport started to catch on. The sport really picked up steam in the mid-1990s, when robot fighting tournaments were first broadcast on cable television. One thing led to another and since then hundreds, if not thousands, of fighting robots have been built. Of these, no two have been built exactly alike.

Some robots, from designers with the artistic leanings of a 1950s Stalinist architect, tend toward the "big steel box" school of robot design. Others borrow a page from Hollywood special effects people, and their robotic creations are so otherworldly they would terrify the giant spiders in *Starship Troopers*.

Most fighting robots have the same basic components. In a nutshell, here is how they work:

Midevil COURTESY TEAM FOAMING RAMPAGE

1. The driver controls his or her robot by giving commands to it via an R/C system. Most, but not all, builders use off-the-shelf or modified FM and PCM radio control systems normally used for controlling R/C vehicles.

2. An R/C radio receiver picks up the radio signals on board the robot. The receiver outputs a tiny and varying voltage; the amount of voltage is based on the signals it receives from the transmitter.

3. The electronic speed control (ESC) interprets the radio receiver's output and, based on that, proportionally controls large amounts of energy and amperage delivered to the motors. Each motor is controlled according to what the radio receiver hears on a particular radio frequency channel. FM radio transmitters can transmit up to eight independent channels. Some computer radio systems can transmit more than three times that many.

4. The motors draw power from the battery according to the control placed on it by the ESC.

5. The rotating motion of the motor is converted to a combat robot usable speed and torque through a drivetrain. A drivetrain consists of components including shafts, gears, pulleys, belts, sprockets, chains, and bearings. Not all of these components are generally used in a single robot. The robot designer picks and chooses the components based on the design.

Here are pictures of the basic fighting robot components:

3.1 Typical DC electric drive motors. COURTESY OF NPC ROBOTICS

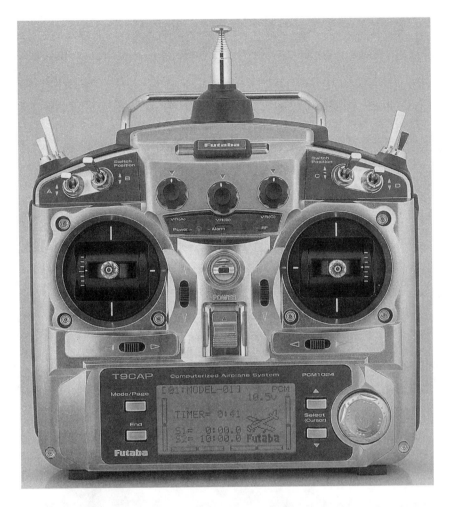

3.2 Radio control systems, operating on legal FM frequencies or in the 900 MHz radio band. COURTESY OF FUTABA CORP.

3.3 Electronic motor speed controlling devices. COURTESY OF VANTEC CORP.

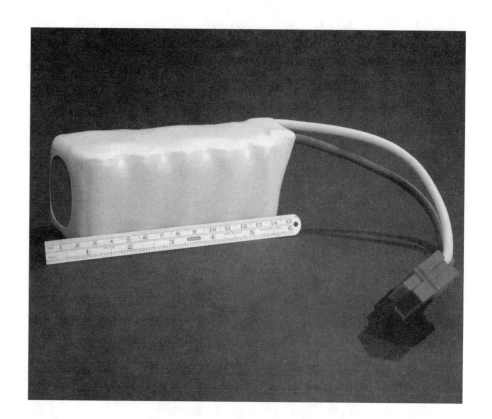

3.4 Rechargeable sealed lead acid (SLA) or NiCad batteries.

3.5 Metal or plastic frames, drivetrains, wheels, and tires.

Add a weapon of your choice, figure out how to connect all of these things together, and, one way or another, you've got a fighting robot.

One common misconception is that the robot builder must be an electronics wizard, a mechanical engineer, and a theoretical physicist all wrapped into one. If you had to build a fighting robot from scratch, you might well have to be all of these things. You'd wind up working 24 hours a day for months on end. But don't despair. The truth is, no one builds everything from scratch. Excellent competitive robots have been built from start to finish in under a month of evening and weekend work.

For instance, most fighting robot builders use the same general type of radio-controlled car and airplane parts to control their robots. No one designs their radio transmitter and receiver systems, their own antennas, and their own electrical interfaces from the radio system to the drive motors. After all, it took teams of engineers at places like Futaba, IFI Robotics, and Airtronics years to engineer the systems that allow you to control your robot reliably, inexpensively, and legally.

Likewise, it takes a multimillion-dollar factory to build the powerful DC motors capable of driving your robot's wheels. If anyone does custom build their own robot motors, I'm unaware of them. Most builders buy their robot motors from a fairly small group of DC motor manufacturers.

There is room for electronics tinkerers to show off some of their hardware building skills. Some very dedicated robot builders have designed their own ESCs to control speed and direction. These folks are in the minority. Most people buy ESCs from firms such as those listed in Appendix I. As builders frequently say, "If you have the option to buy a part or make a part—just buy it."

The chapters that follow will examine the components that all fighting robots have: motors, ESCs, radio systems, mechanical parts, and weapon systems. If you are truly an innovator, take heart. There is plenty of room for innovation. Individuals have broken and continue to break every so-called rule in this book. Sometimes it works, and sometimes it doesn't. If you want to follow the general methods for building bots, that's great. If you

3.6 Parts of a typical fighting robot

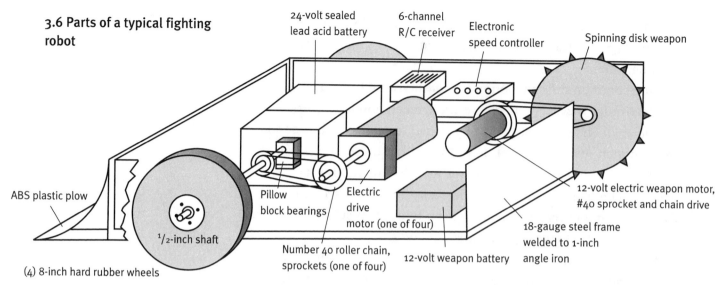

24-volt sealed lead acid battery

6-channel R/C receiver

Electronic speed controller

Spinning disk weapon

ABS plastic plow

Pillow block bearings

Electric drive motor (one of four)

$^1/_2$-inch shaft

12-volt electric weapon motor, #40 sprocket and chain drive

(4) 8-inch hard rubber wheels

Number 40 roller chain, sprockets (one of four)

12-volt weapon battery

18-gauge steel frame welded to 1-inch angle iron

Other tires and motors omitted for clarity

3.7 Fighting robot schematic drawing

Most robots use similar components:
"A" Electronic speed controller
"B" Wiring termination strip. All electrical components are wired to this point and cross-connected.
"C" Radio signal receiver
"D" Drive motors
"E" Drivetrain: bearings, shafts, roller chain, sprocket, wheels
"F" Drive motor batteries
"G" Weapon: A motor-driven spinning disk
"H" Weapon motor battery
"I" Robot frame

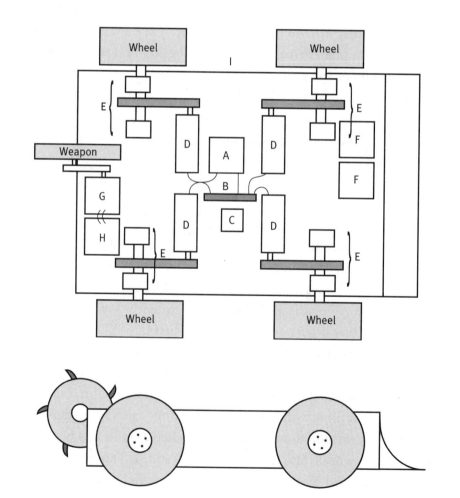

3.8 Side view of a fighting robot

march to a different, robot drummer—if you've decided to use a big rubber band to power your robot, or use shouted voice commands instead of radio control—go for it. Your innovation may signal the next big thing in the sport. In the diagrams to the left, each main section of a fighting robot is identified. You'll find detailed descriptions of every subsection in the chapters that follow.

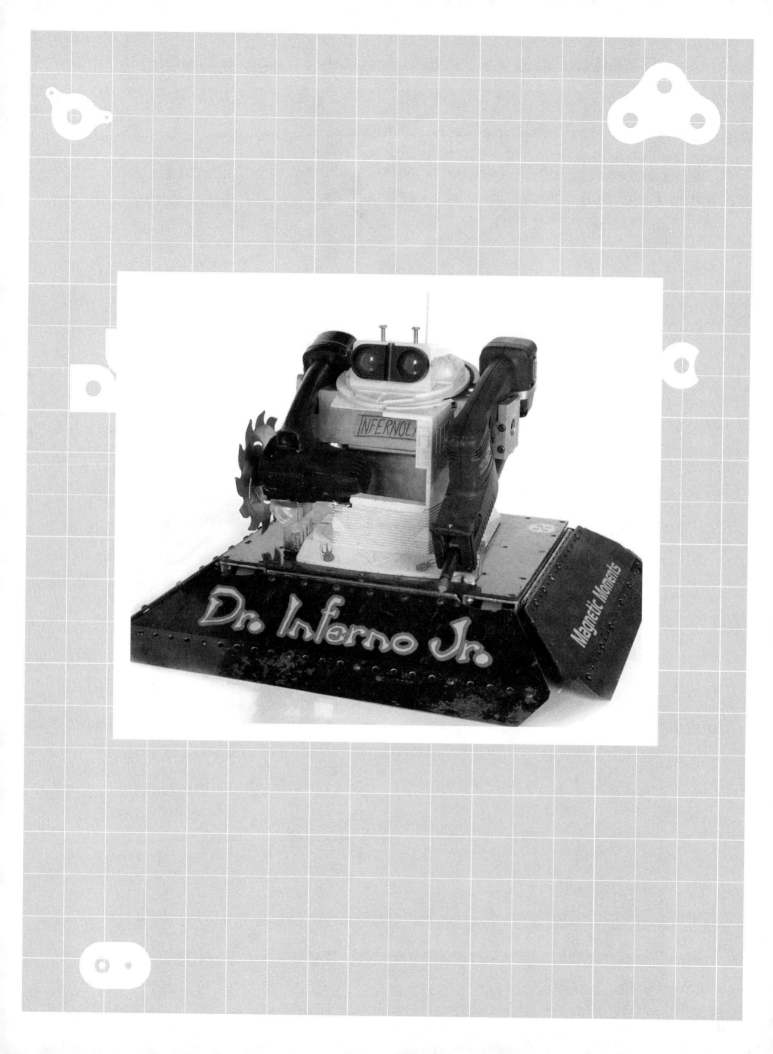

4.
RADIO CONTROL SYSTEMS

Combat robots are steered by remote controlled radio systems, generally referred to simply as R/C. Most robots builders use the radio-control equipment typically associated with R/C airplanes, cars, and boats, and modify it slightly to control the fighting bot. There are others ways to control a robot, such as by sound, a tether wire, or infrared devices, but most competitions require the use of either standardized R/C controllers or newer 900 MHz computer radio telemetry.

The combat robot's radio system consists of three main parts. First, there is the transmitter, the squarish box or gun handle with a telescoping antenna that you hold in your hand. Next there is the receiver, a small box with wire connections to the robot's motor speed controllers, relays, and solenoids. Finally, there is the battery pack, which is a small rechargeable NiCad battery pack. The radio-controlled vehicle industry is large, and the vehicles change frequently. As a robot builder, you can learn a lot by talking to members of your local R/C flying and car clubs.

The Basics

In the United States, there are a number of frequency ranges, or channels, available for use by robot builders,

Dr Inferno, Jr COURTESY
BATTLEBOTS, INC., AND JASON DANTE BARDIS

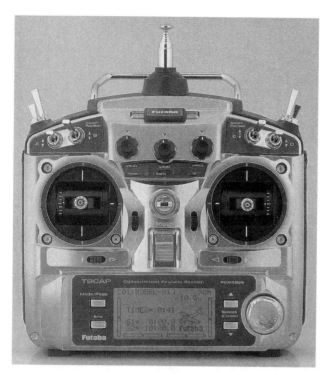

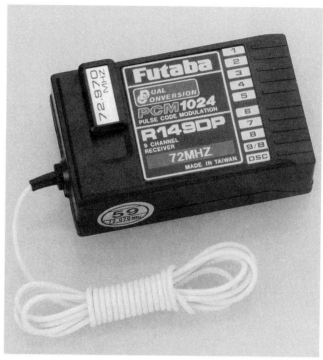

4.1 Radio transmitter. COURTESY OF FUTABA CORP.

4.2 Radio receiver. COURTESY OF FUTABA CORP.

model airplane flyers, and model car builders. Specifically, the parts of the radio spectrum allocated by the Federal Communications Commission for unlicensed radio control devices include the 27 MHz land/air band that is used for inexpensive or toy systems, the large 72 MHz band for flying machines, and the 75 MHz band for terrestrial systems. While it may seem like there are a lot of available frequencies, there are usually more—sometimes far more—competitors in big tournaments than there are available frequencies.

Most serious U.S. robot builders currently stay in the 75 MHz (called the "land only") band. The 27 MHz band is really for toys, and the 72 MHz band is for model flying machines like airplanes and helicopters. The rules for some competitions require that competitors use only the 75 MHz band. Those same tournaments may also require the use of PCM-FM radio equipment. (PCM, or "pulse code modulation," is explained later in this chapter.) In the recent past, no standard PCM radio in the 75 MHz frequency range was available off the shelf. So builders were required to order a special modification of PCM equipment by the manufacturer to work in the 75 MHz range. Also, in the United Kingdom, government regulatory agencies have allocated a different portion of the electromagnetic spectrum to R/C models. It is called the 40 MHz frequency and is solely dedicated to land-based models. This frequency allocation provides for 34 channels, each with a 10 kHz channel spacing. The center frequency of the first channel is 40.665 MHz.

The roboteer controls a combat robot over a radio link using a specific frequency. In any meet or tournament, great care is usually taken to ensure that everybody knows which bot on the competition floor is controlled by which frequency. Since there are only a limited number of radio frequencies available and often a great number of robots participating in a meet, the event's sponsors will normally have stringent safety procedures in place to limit what transmitters are broadcasting at any particular time. *Always* follow the rules regarding turning your radio on and off. Causing a 12-horsepower, 300-pound robot with a metal cutting saw to turn on unexpectedly could be tragic!

AM and FM Radio

Bots are controlled using the same R/C equipment that model airplane and car builders have used for years. The biggest radio manufacturers are Airtronics, Futaba, Hitec, and JR. There are other, smaller firms besides these large companies. Most manufacturers make similar products; however, each company has its own ardent supporters. Most of these companies make several types of radio systems. The two most common types of radio systems that robot builders can use are called AM and FM.

AM Radio Control

AM stands for amplitude modulation, and it is the oldest and cheapest form of radio control. An AM radio system consists of a handheld radio transmitter and a radio receiver that is mounted near the motors on the robot. When the AM radio user turns the transmitter on, it begins to transmit a low-power radio signal. In AM radio the strength, or amplitude, of the radio signal is modulated by the information being sent by the user's transmitter. A word of advice here: AM modulation is really a nightmare in metal banging, electrically powered robots, and most builders would not recommend it. It is a starting point for guys with old radios in their garages. After losing their hair, they move on to something better pretty quickly.

FM Radio Control

Frequency modulation, or FM, transmits a constant strength, or amplitude, radio wave. But the frequency changes from slow to fast and back again depending on the information being sent by the user. FM has many advantages over AM. It is more reliable, less sensitive to environmental disturbances, and received with less distortion. In fact, most combat robot tournaments don't allow the use of AM radio control systems. AM is often considered too prone to interference and loss of control compared to FM systems. No one wants an uncontrollable robot on a rampage.

Regardless of whether you choose to use AM or FM, both radio systems work similarly. The information to be transmitted is "impressed" on the AM or FM carrier wave and transmitted across the competition floor to the receiver inside the robot. There, the electronics inside the receiver process and extract the information sent by the robot's driver. What kind of information? Specifically, the information about the position of the driving stick, the position of the throttle, the position of the weapons, and so forth. Inside the transmitter, special electronic components combine two different types of radio signals and send the modulated signal out to your robot. The first signal is the radio frequency itself, unique to your robot, which broadcasts the information out to the battle floor. This signal is called the "carrier signal." The second signal is a series of encoded pulses that do the actual work of telling the robot what direction to go, how fast, and so forth. The radio frequency drawings show how the carrier wave is combined with the information pulses to make a controlling radio signal.

The radio receiver mounted inside your robot hears the carrier signal and the radio signals from all the other drivers' transmitters and other radio stations too. But each receiver has a special tuning device inside it called a crystal. A crystal is usually made from mineral quartz and can be tuned to work only with radio energy in a very narrow range. Each crystal is made so that it allows robot control information from one, and only one, frequency to pass from the antenna through it into the internal decoder of the receiver. When the radio signal reaches the radio receiver mounted within the robot, the crystal filters out the signals belonging to the other radio frequency users and sends through only your own robot control commands—go left, speed up, slow down—from the driver.

Now it's time to explore how the information from your stick or throttle on the transmitter gets sent and understood by your robot. The first concept to understand is something called pulse position modulation, or PPM.

Pulse Position Modulation

Imagine you've just built a very simple robot, with a single motor, and that motor can make the robot move in a straight line, forward or backward. Whether it goes forward or backward depends on whether you push the control stick up (forward) or down (backward). Call this robot Borebot, because this is not a very interesting robot!

To control Borebot, you need a single radio channel. Say that your radio has a crystal in it that allows it to operate on the 75.970 MHz frequency. When you turn your radio on, your transmitter will sense the position of the joystick and constantly send out position information on the 75.970 MHz frequency channel on your radio. (The radio frequencies available in the United States are included in Appendix II.)

Most robot control systems are based on the control system first used to fly model airplanes. Model airplane systems use something called pulse position modulation (PPM) to set flaps, throttles, and rudders. Although the details of radio control differ among manufacturers, this simplified version of PPM will give you the general idea. If you want your robot to move forward, you pull the joystick up. Then, the transmitter reads the full up command and the electronics within it send out radio pulses of duration 2.0 milliseconds, once every 20 milliseconds, as shown in Diagram 4.3. The motor speed controls are connected by wires to the radio receiver on your robot (covered in Chapter 7), and the controller is programmed so that a 2.0 millisecond voltage pulse from the receiver means "go forward." So it allows current to enter the motor windings and the motor rotates forward. If the stick is held halfway between "go forward" and "go backward" (that is, stay idle) the transmitter likewise reads the stick position and sends out a pulse of 1.5 milliseconds every 20 milliseconds. The motor speed controller also knows that a 1.5 millisecond pulse means "stay idle," so Borebot just sits there. If the stick is placed in the full

4.3 Single channel radio system block diagram

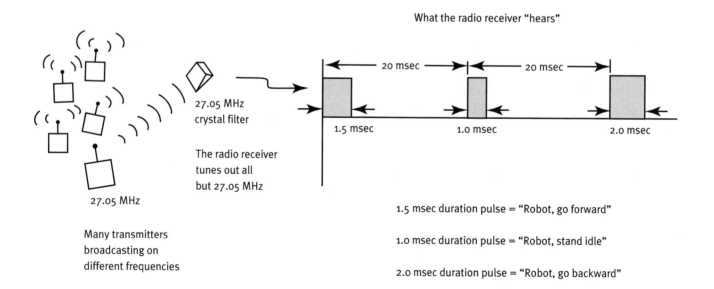

What the radio receiver "hears"

20 msec   20 msec

1.5 msec   1.0 msec   2.0 msec

27.05 MHz crystal filter

The radio receiver tunes out all but 27.05 MHz

27.05 MHz

Many transmitters broadcasting on different frequencies

1.5 msec duration pulse = "Robot, go forward"

1.0 msec duration pulse = "Robot, stand idle"

2.0 msec duration pulse = "Robot, go backward"

backup position, a pulse of 1.0 millisecond duration is sent out every 20 milliseconds. As before, the speed controllers are aware of what a 1.0 millisecond voltage pulse means and send the appropriate "full speed backward" voltage to the motor.

In summary, it is the duration of the radio pulse, ranging between 1.0 and 2.0 milliseconds and sent out every 20 milliseconds, that the robot radio receiver hears and interprets. This control scheme, based on a series of repeating pulses of lengths varying between 1.0 and 2.0 milliseconds every 20 milliseconds, is called PPM.

Now consider how to control a robot with more than one motor. You can see in Diagram 4.4 how the signal might look if you hooked up a device called an oscilloscope to the output of the radio receiver. An oscilloscope traces a voltage on a screen over time. A lot of time is wasted during that 20 millisecond cycle, just waiting for the next pulse. So add another motor to Borebot and control a second motor. To control the second motor you will instruct the radio receiver to wait about 2.5 milliseconds after the start of the first pulse and take whatever shows up in this time slot as pertaining to the second motor. This is also shown in the radio control block diagram in Diagram 4.4.

After the transmitter is turned on, it immediately begins to send out a pulse of duration between 1.0 and 2.0 milliseconds. This pulse corresponds to the position of joystick one and therefore controls motor number one. Two and one-half milliseconds after the first pulse, the transmitter sends a second pulse, again lasting between 1.0 and 2.0 milliseconds. The second pulse in the sequence corresponds to the position of joystick two and controls motor number two. Now Borebot can turn left and right and is much less boring.

The whole sequence of pulse one, wait, pulse two, wait, et cetera, repeats every 20 milliseconds. The duration and position of the pulse within the 20 millisecond timing cycle controls all of the motors and other devices on the bot. As you can see in Diagram 4.4, you have room within the 20 millisecond cycle to place up to eight pulses, all 2.0 milliseconds long with a half-millisecond in between. Therefore, you can control up to eight different devices

In this simplified R/C control scheme, a "frame" of information is sent every 20 milliseconds. The R/C equipment knows that the length of the pulse during the first 2.0 milliseconds corresponds to Channel 1, there will be a 0.5 millisecond quiet period between channels, and then the Channel 2 information will be sent. So on for Channels 3 through 8.

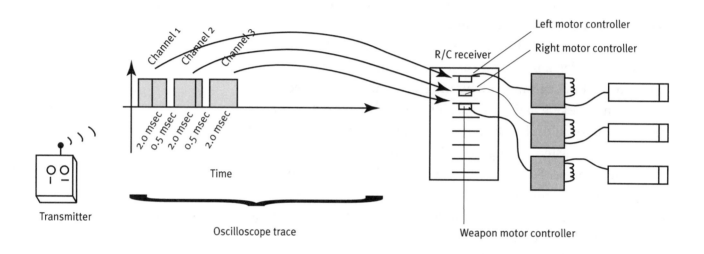

4.4 Multiple channel radio system block diagram

from one transmitter on one channel. This ability to place up to eight pulses within a standard 20 millisecond cycle, all controlling different things, allows the roboteer to control multiple motors, weapons, lights, and so forth from a single transmitter using a single frequency.

Pulse Code Modulation

Next, consider another radio control scheme called pulse code modulation, or PCM, systems. Many builders and R/C hobbyists consider PCM a step up from PPM in terms of sophistication. Pulse position modulation, when used to control fighting robots, is really just an adaptation of the control scheme R/C airplane pilots have been using to control flying airplane throttle and rudder positioning devices, called servo motors. Robot builders can and do use PPM to control robots, and most of the time it works pretty well. However, PPM requires that a stable 20 millisecond frame be set up and information be sent at very precise intervals in order to control the robot. If a timing, or sequencing, error occurs, the battling bot doesn't know what to do, at least until it is resynchronized. This is termed a "glitch." While the glitch lasts, the robot usually continues to do what it was doing immediately before the glitch. If it was going straight, it still goes straight. If it was turning, it continues to turn. Some tournament sponsors see this as a safety hazard.

PCM is an alternative to PPM. Instead of sending 1.0 to 2.0 millisecond pulse information within a timed frame, PCM simply sends a stream of digital information directly over the FM radio wave to the receiver. The receiver recognizes the digital data stream and breaks it down into header information, an error checking section, and the control information itself. A microcomputer inside the receiver decodes the information and then tells the electronic motor controller what to do. PCM is fast, clean, and sure. If a glitch occurs, there is a microcomputer right on board the receiver, so the robot can possibly be programmed to stop. This is why many folks feel it is safer than PPM. A disadvantage is that each radio manufacturer has its own version, so a PCM transmitter from one manufacturer won't work with a PCM receiver from another. In the PPM world, everybody's equipment generally works with at least somebody else's, so it's easier to get replacement parts and the prices are much more competitive.

900 Megahertz Systems

900 Megahertz (MHz) telemetry systems are different from standard AM and FM systems in several ways. They use a complicated radio technology called spread spectrum frequency allocation instead of AM or FM broadcasting. It is interesting to

note that the idea for spread spectrum was first formulated and patented by the famous actress Hedy Lamarr. In 1942, she and American composer George Antheil developed and patented the idea for the technology after being inspired by examining the holes in a player piano roll.

The spread spectrum technology used in fighting robots utilizes a band of radio frequency spectrum at the 900 MHz level in which the Federal Communication Commission (FCC) allows unlicensed, low-power radio communications. Many wireless phones use this bandwidth. Instead of allocating exact frequencies to individuals at tournaments, everyone uses the same frequency at the same time. How can this be? The devices use sophisticated telecommunications circuitry to digitally encode identity and address information on the signals transmitted. Each robot is listening to the 900 MHz broadcast but screens out everything except the information destined for it alone. The information is encoded, and it hops between frequencies within the 900 MHz band using bandwidth as it is available.

There are several key differences between 900 MHz systems and 75 MHz systems. First, with 900 MHz, the robot receiver and the robot transmitter engage in a two-way conversation. Therefore, information about the status of various components and systems onboard the robot can be relayed back to the driver. For example, the robot's battery life can be displayed on the driver's controller.

Second, and perhaps more important, organizers of robotic competitions always worry, and rightfully so, that someone is going to turn on a radio transmitter when they shouldn't and start up somebody else's robot by accident. This could lead to tragic results. There are only a relatively small number of channels available in the 72 or 75 MHz radio band, and at big events, several different groups will necessarily be using the same frequency. Such problems don't arise with computerized controllers because the information is specifically encoded for individual robots.

Another advantage of 900 MHz over lower frequency radio systems is that the line of sight transmission range is farther for the same amount of radio power. Therefore, the robot can wander farther away and still be controlled by your radio transmitter. On the other hand, the 72 or 75 MHz radio waves are better able to penetrate through objects. So, if another robot or a part of the arena hides your robot's antenna, the lower frequency systems often have an advantage.

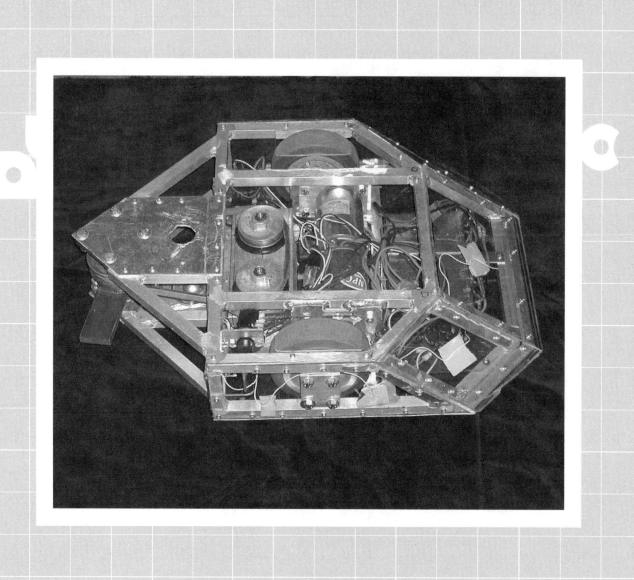

5.
ELECTRIC MOTORS

There are several methods for making your robot move. You could use an internal combustion engine, a hydraulic motor, or even a controlled jet of pressurized air. But the vast majority of robot builders use direct-current electric motors to power their bots. If this is your first fighting robot, you should use a direct current, or DC, motor. Here's why:

1. They're widely available.
2. Their speed is easily controlled.
3. It is simple to devise a control system that reverses direction.
4. Batteries to power them are widely available, and, unlike gasoline motors, they are generally safe to use in a battle.
5. They have a decent power to weight ratio.

Some builders use internal combustion, or IC, engines. Internal combustion engines provide a lot of power for their weight, but they increase the complexity of your bot design. Consider how long it could take and how expensive it is to design a workable system to reverse the wheel direction, a safety fuel cutoff system, and an IC engine speed controller, and you'll likely do what most bot builders do: go electric, at least for driving motors.

This being said, of all the critical parts on your robot, none is more important (or probably as expensive) as the

Detassler COURTESY TEAM DOS

45

DC electric motors. After all, when the robot stops moving, it's done and you're done. This chapter provides the robot builder with the information to choose and use the motors needed to make a competitive fighting robot.

Motor Basics

What is a motor? A motor is simply a machine that converts electrical energy into mechanical motion. Motors work by using electro-magnetism to impart a rotation to a shaft. The magnetically driven spinning shaft is then connected to wheels, and voilà! You've got motion!

Assuming you're new to building fighting bots, there are a few things to know before you go to the surplus store or surf the Internet and start buying electrical motors. First of all, motors can be powered by alternating current (AC) or direct current (DC). Some types of motors called "universal motors" run on AC or DC. But generally, universal motors require high DC voltages to operate efficiently, so they are not very common in fighting robot tournaments. For the most part, AC and DC motors are constructed differently and are not interchangeable. So, unless you plan to outfit your bot with a portable AC generator or a very, very long extension cord, you won't have much interest in AC motors.

The next thing to be aware of is that there are different types of DC motors. Not all of them are suitable for this type of use. For example, you might come across DC motors called "stepping motors." These motors are designed to turn a repeatable amount every time current is applied to an input lead. Stepping motors are designed to precisely position objects. They're quite commonly offered for sale, but unmodified they are not much use for moving robots around the battle floor.

What bot builders are interested in is a DC, open loop, continuous motor. Inside the motor, there is a rotating element called an armature. The armature, sometimes called a rotor, is crisscrossed with copper wire, or "windings," through which electric current flows. The electromagnetic fields produced when

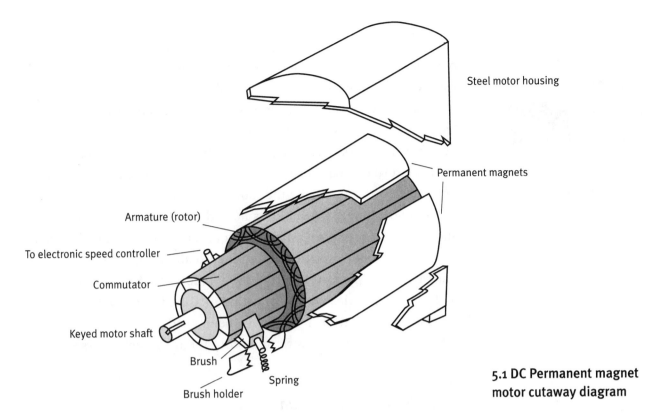

Steel motor housing

Permanent magnets

Armature (rotor)

To electronic speed controller

Commutator

Keyed motor shaft

Brush

Spring

Brush holder

5.1 DC Permanent magnet motor cutaway diagram

current flows through the armature windings interact with alternating electromagnetic fields set up in the field windings in the motor housing (called the stator, because they are stationary). The interaction is complex, but basically, an unending cycle of attracting and repelling electromagnetic fields is induced in the rotor and stator windings. This causes the rotor to spin on and on, chasing but never catching the field windings, just like a greyhound chasing a mechanical rabbit.

There are three different types of DC open loop motors. All DC motors have an armature, field windings, and a housing. How the field windings are set up changes the operating characteristics of the motor. The three ways to set up field windings are (1) to use permanent magnets, (2) to use series wound electromagnets, or (3) to use parallel or "shunt" wound electromagnets. In a permanent magnet DC motor, permanent magnets line the inside of the motor housing. The armature is made from a hunk of iron wrapped in multiple turns of copper wire. In series and shunt wound motors, the field windings are electromagnets, wired to the same power source as the armature. Each of the three methods of DC motor winding has its advantages and disadvantages. In general,

5.2 DC electric motor

permanent magnet and shunt motors are more easily adapted to use as the main drive motors for a robot than are series wound motors. The problem with series wound motors is that they are harder to reverse. While permanent magnet and shunt wound motors can be easily reversed by simply switching the direction of the armature current, a series wound motor must be completely rewired in order to enable the motor to reverse.

Many beginning bot builders head down to the local junkyard and purchase a couple of starter motors, assuming that any low-cost, 12-volt DC motor capable of spinning up an automobile engine must be really powerful and torquey. True enough, but starter motors, being series wound, must be modified to make them reverse. While starter motors have been used successfully to drive robots, it takes a lot of work to modify them so they reverse.

When to Use Series Wound Motors

A series wound motor has similar performance characteristics to a permanent magnet DC motor, at least when it comes to examining

5.3 High-torque weapon motor. COURTESY NPC ROBOTICS

motor data sheet specifications. However, there are differences in terms of what happens within the motor as the speed of the motor increases. If you remove the load from a series motor, it turns faster, the voltage increases, and the torque and the current drawn decreases. (Torque is the turning force that a motor puts out.)

This is unlike a permanent magnet motor, where the stator, or outer field, strength is always the same. The permanent magnets have just so much of a magnetic field and that's it. But in a series wound motor, the stator field comes from electromagnets, not permanent magnets. You'll find that, as you take load off the motor, the series motor turns faster. This is because as the motor turns faster, the magnetic field in the electromagnets becomes weaker. Next, as the motor spins faster, and the magnetic field in the stator weakens even more and the armature begins to spin even faster. So, series wound motors tend to rev to very high RPMs. Some junky series wound motors will self-destruct if run unloaded; they just fling themselves apart. On the other hand, a series wound motor will, for the same reasons, put out lots of torque at high motor loadings.

These characteristics—high stall torques and high unloaded motor speeds—make series wound motors desirable for some applications involving spinning robot weapons. Spinning weapons are covered in more detail in Chapter 12, but these types of weapons require a fast spin-up time and high top-end rotational speed, and that's just what series wound motors deliver.

The bottom line on series wound motors is: good for many spinning weapons, but not so good for most robot main drives.

DC Brushless Motors

DC brushless motors are the latest development in electric motors and have been used very successfully by some robot builders. The DC brushless motors are built quite differently from conventional DC motors. Instead of using carbon brushes to make alternating contact with the armature, brushless motors use an electronic sensor to track the precise position of the spinning armature. This information is relayed to a specialized, computerized control circuit that pulses energy into coils lining the inside of the motor. The computer optimizes the timing of the pulses and the powerful armature magnets make DC brushless motors very efficient, powerful, and relatively small. (The famous Segway or "Ginger" electronic people mover uses brushless DC motors.)

How to Choose a Motor

In practice, most builders don't choose their motor based on the type of internal motor wiring. Instead, they first decide what they want their motor to do. Then they check out the various motors available on the Internet or at the surplus store to see what will meet their requirements. The robot designer does this by examining a document called a motor data sheet. A good motor data sheet provides the critical pieces of information in a graphic or table format. The information is based on a dynamometer test of the motor at various loadings.

MOTOR DATA SHEET

Dynamometer test results

Torque in inch-pounds	Amps	RPM	HP
30	8.6	238	.11
60	12.5	230	.22
90	16.2	225	.32
120	20.0	218	.41
150	23.5	211	.52
180	27.5	206	.62
210	31.6	200	.71
240	35.1	194	.81
270	39.2	187	.89
300	43.1	181	.95
825	110	STALL	STALL

5.4 Typical motor data sheet

The data sheet columns provide the following information:

1. Torque at the rotor output
2. Amperage drawn at various torques
3. The rotor speed in revolutions per minute at various torque levels
4. The horsepower rating of the motor at various torque levels

The motor data sheet is an important document. It provides the information you need to choose the correct electronic speed controller and battery as well as the motor itself. The data sheet will tell you the strength with which the motor can turn a load under various conditions. The torque put out by a motor at any particular instant depends on the size of the load it is asked to move. When your robot is pushing hard against an opponent, your motor is putting out a lot of torque. When it's just cruising from one side of the arena to the other, it isn't putting out much torque at all.

Permanent magnet and shunt wired motors have a particular level of loading, called the motor design load, where the motor develops its maximum power. Under lighter or heavier loading conditions, the amount of power put out by the motor is less because either the torque is smaller (load is lighter) or the motor slows down when the load increases past the design speed. Don't misinterpret this statement and think that a heavy torque won't burn out a permanent magnet motor's field windings or the electronic motor controller just because the motor slows down. Also, a load too heavy for the motor's design can crack motor magnets as well. As you can see from the typical motor data sheet, the amperage drawn continues to go up with the amount of loading. There is a linear relationship between the torque (load) and the amps drawn, but not between the horsepower developed and the amps drawn. Too big a load for too long a time will fry your motor no matter what kind it is.

Combating Radio Frequency Noise

We discussed the differences between brushed and brushless motors earlier. Brushless motors have some advantages in terms of size and power density. But most builders use brush DC motors because they cost less and the control equipment is simpler. (The brushes are devices designed to make periodic connectivity between the internal field windings and the spinning armature.) However, the on/off action of the brushes can cause a side effect, called RF noise, that can interfere with your radio control. To counteract this problem, many builders place a small capacitor between the external brush connection and the motor housing. In many cases, placing a 1 to 10 microfarad ceramic capacitor between each lead on the motor and the motor case does the trick.

Torque

Start by figuring out how much torque you need. This relates back to the original design decisions you made when you first dreamed up your fighting robot. Will your bot be a brute force sort of machine, trying to bull other robots into the arena's hazards? If so, you will need a *lot* of motor torque. On the other hand, maybe you want your bot to be basically a platform for spinning or kinetic energy weapons. If so, then you might need only so much torque for the drive motors, but as much torque as you can get for the weapon's power motor.

Revolutions per Minute (RPM)

Next determine out how fast you want your motor to turn. Most motors are designed to run at speeds far higher than could be reasonably controlled during a competition. And, among the many trade-offs out there is speed versus torque. A fast-turning motor will have little torque or pushing power. Therefore, most motors are connected to the wheels via a drivetrain of belts, chains, or gears, to reduce speeds and increase torque. You get speed or RPM data from the motor data sheet. Once you know the speed characteristics of the motor, go to Chapter 9. There, you'll learn how to use sprockets, chains, and gears to obtain the desired bot velocity from the motor's design speed. The following chart gives some idea of how fast a robot will go given the size of its drive wheels and their

RPM. For example, to move a robot with 6-inch wheels along at a fast 9 MPH clip, the motor has to turn them at a rate of 500 RPM.

SPEED–RPM–WHEEL DIAMETER RELATIONSHIPS

Wheel Diameter (inches)	Revolutions per Minute	Forward Speed (MPH)
3	100	0.9
3	300	2.7
3	500	4.5
3	1000	8.9
4	100	1.2
4	300	3.6
4	500	5.9
4	1000	11.9
5	100	1.5
5	300	4.5
5	500	7.4
5	1000	14.9
6	100	1.8
6	300	5.4
6	500	8.9
6	1000	17.8
8	100	2.4
8	300	7.1
8	500	11.9
8	1000	23.8

For beginning builders, it is possible to connect a motor directly to a wheel without any sort of drivetrain. If you plan to do this, then you will have to shop for a special type of motor called a gear head motor, or simply a gear motor that has the necessary mechanical speed reduction equipment built right into it. You can look for a gear head motor with a design speed rating of between 150 and 400 RPM. Then, by choosing the appropriate wheel diameter, the clever novice bot builder can much more easily build a robot with a operating fighting speed of say, 6 to 10 miles per hour without any additional gears or sprockets. Several suppliers in the motor section of Appendix I specialize in supplying gear head motors for robots.

Amperage

Finally, check the motor data sheet for the amperage rating at various torque loads. You'll need to make or buy an electronic speed controller for your robot that can handle the rated loads without burning up. (See Chapter 7 on motor speed controllers.) The motor will demand the highest amperage at stall condition, when the rotor is locked up. The motor specification charts will tell you the current draw during the following four phases of any combat match: (1) at stall, (2) while pushing, (3) while accelerating, and (4) while moving at constant velocity. Knowing the maximum power draw (during stall) tells you how big an electronic speed controller you need.

Voltage

Electrical motors are not 100 percent efficient. The losses are due to many causes such as friction and resistance. The copper windings of a motor do have a small yet significant resistance. This equation computes the losses in a motor due to resistance in the motor windings:

Loss (watts) = Current² (amps) x Winding Resistance (ohms)

This equation explains why high-voltage motors are more efficient than low-voltage motors. Since the resistive losses are calculated from the square of the current multiplied by the motor's resistance, a high-voltage motor will have far less loss. A 10-horsepower motor run at 6 volts will draw 2,500 amps at peak power, but a 10-horsepower motor run at 60 volts will draw only 250 amps at peak power. In general, the higher the motor voltage is, the more efficient the motor.

It is a time-honored tradition in robot building to use more voltage than the motor manufacturer's data sheet specifies. If you run the motor at less voltage, the motor goes slower and produces less torque. If you run it at higher voltage, it produces more speed and will have more torque. Every motor is different, but many builders routinely run their robot motors at higher than the rated voltage.

If you double the voltage, you more than double the current that the motor could draw. Note that I say "could draw," not "will draw." Your battery must be able to output that much current, and your motor's electronic speed controller must be able to handle it as well. If both electronic speed controller and battery are able to handle it, you will get four times the horsepower by doubling the voltage. The RPM of the motor will also be doubled. This sounds pretty good—more power, more speed—but remember that robot building is all about trade-offs. The motor will also get four times as hot, use up four times as much battery capacity, and probably wear out four times faster. That is the main problem with over-volting your motors. Also, be aware that some motors are built so they cannot be safely overvoltaged. Even a 10 percent overvoltage will cause those motors to snap, crackle, and pop. Other motors, built differently and more ruggedly, will put up with this sort of abuse without too much complaining. Each motor is different, and it's tough to know in advance unless a robot builder friend with a similar motor has tried it out beforehand. In general, motors that are large and heavy for their output tend to overvoltage well and, conversely, smaller, lightweight motors for a given power output do not overvolt as well. For motor suppliers, see Appendix I.

Horsepower, Voltage, and Current

It is important to understand the meaning of motor horsepower. Everyone knows that the higher the horsepower rating, the more powerful the motor. This is both good and bad, as you'll see.

First, the good news about horsepower. With good gearing and traction, higher horsepower ratings provide you with a greater ability to push other robots around the arena. If robot design is basically just a platform with batteries, motors, and wheels, then by all means get as much horsepower as you can handle. But "as much as you can handle" leads us to the bad news. The horsepower rating of an electric motor is proportional to the product of the motor's amperage draw at stall condition and its voltage. Your voltage is pretty much determined by the upper limit that your motors can handle. We talked earlier in this chapter about overvoltaging your motors.

For a given voltage, the more horsepower developed, the higher the amperage within the motor windings. The thing is, high current is hard to handle. Why is that? Well, high current requires large diameter wires throughout your bot. Wire is cheap, so that's no real issue. High current also requires on/off switches, terminal strips, and other miscellaneous equipment rated to meet those requirements. It is a little more expensive, but still doable. High current requires motor control devices that can accommodate high amperage, usually called electronic speed controllers (ESCs). This is one of the rubs: ESCs are expensive to buy. Large ESCs are *very* expensive. Take a very large ESC capable of controlling four 2-horsepower motors, as an example. Eight horsepower requires an ESC capable of handling very high amperage.

The relationship between volts, amps, and horsepower is:

$$\frac{\text{Voltage (volts) x Current (amps) x Efficiency}}{746} = \text{Power (hp)}$$

or

$$\text{Current (amps)} = \frac{\text{Power (hp) x 746}}{\text{Voltage (volts) x Efficiency}}$$

So, if you have a 24-volt system of motors putting out eight horsepower, and assuming 50 percent efficiency, the current draw is:

$$\text{Current} = \frac{8 \text{ hp x 746}}{24 \text{ volts x 50\%}} = \text{about 500 amps}$$

That's a ton of amperage! Few ESCs can handle this much current for a sustained period of time. Obviously some planning is necessary in order to make it manageable. We'll discuss how to select ESCs in more detail in Chapter 7. The other problem with high motor currents is the internal motor heating that occurs. High currents cause the motor to heat up, which could have a very bad effect on its life. Some builders counteract this by adding a fan to cool the motors. Blowing air over the motors will keep them cooler. However, this technique does add weight and decreases battery capacity since the fans draw power as well. A better solution is to engineer a heat sink into the motor mounting to "wick away" and dissipate the excess heat.

No Motor Data Sheet?

Sometimes bot builders just want to use what they already have. For instance, imagine that your Uncle Bob just happens to have two motors in his garage and is anxious for you to take them because he thinks they're junk. From the nameplate on the motors you can see they are 24-volt DC motors, but there's no other information. What do you do now?

You test them! You need to figure out how to find the big four quantities of robot motordom: how much torque they put out, how much amperage they draw, how fast they turn, and the maximum horsepower they produce.

First, test them for torque. You can make a pretty accurate assessment of the mystery motor's torque by constructing something using a spring scale with a one-foot lever arm attached to the motor shaft. From that, you can read the torque in foot-pounds directly off the spring scale. You'll know just one parameter—the stall torque of the motor—but this is a pretty good thing to know. Stall torque occurs when current is applied to the motor, but the shaft is held immobile. Don't do this for more than a second or two, as it can fry your motor, your motor controller, your battery, and everything else. There will be times during the heat of battle when you need to apply a lot of current, but minimize such occurrences whenever possible.

When your robot is pushing as hard as it can against another bot and nothing out there is moving, it's putting out its maximum, or stall, torque. In general, the more you can push, the better.

Now you need to test the motor for its maximum rotational speed. Finding the maximum motor speed is trickier. The easiest way to figure the rate at which the motor shaft is turning is with a tachometer. Tachometers are fairly inexpensive (less than 30 dollars) and available from a variety of places including model airplane supply houses. They are designed for measuring the rotational speed of a propeller, but it isn't too hard to adapt them for robot use. You simply place a protrusion connected to the turning motor shaft between the tach and the sun and read the speed right off the tach's dial. Without a digital tachometer, the question is trickier to answer. One solution is to measure the amount of time it takes the motor with a pulley on it to reel in a measured length of rope attached to the motor shaft. If you know the length of the rope, the diameter of the pulley, and the amount of time it takes to reel it in, you can determine the rotational speed from the following equation:

$$\text{Length of rope (in) x } \frac{1 \text{ revolution}}{\text{Pulley diameter (in) x } \pi} \text{ x } \frac{1}{\text{Time (sec)}}$$

$$\text{x } \frac{60 \text{ sec}}{\text{minute}} = \frac{\text{revolutions}}{\text{minute}}$$

For example, imagine that you placed a 5-inch diameter pulley on the motor shaft and attached a 120-inch rope to it. You timed the period it took to completely wind the string to be 5 seconds. Then the speed of the motor is:

$$120\text{-inch rope x } \frac{1 \text{ revolution}}{5 \text{ inches x } 3.14} \text{ x } \frac{1}{5 \text{ seconds}} \text{ x } \frac{60 \text{ seconds}}{\text{minute}} = 92 \text{ RPM}$$

Horsepower

Now you know the no-load RPM from the string test and the stall torque from the scale test. From these two parameters, you can obtain a pretty close approximation (plus or minus 25 percent) of the peak horsepower developed by the motor by using this formula:

$$\text{Peak Mystery Motor hp} = \frac{0.5 \text{ Torque at stall (foot-pounds) x } 0.5 \text{ RPM no load}}{5,252}$$

Finally, you need to test for the maximum current drain, or the amount of current flowing through the motor at stall torque. The best way to do this is with an ammeter made for this purpose. Ammeters are readily available at many electronics stores and radio-control hobby stores.

Wiring Basics

Robots have quite a bit of wiring in them. It is important to use the correct wire size for the job at hand. If you wire your robot with too thin a wire, the wire could melt. If you use too thick a wire, you add needless weight. Wire insulation may be made from PVC plastic or it may be made from non-PVC materials, such as Teflon or Kepton. Some builders prefer non-PVC insulation as it is slightly lighter in weight, although it does cost considerably more.

The diameter of electrical wire is expressed in terms of its gauge. The lower the gauge number, the thicker the diameter of the wire. The diameter of the wire and the amount of insulation surrounding the wire determine how much current that wire can safely carry. This chart shows the number of amps a wire of a given gauge or diameter can handle.

MAXIMUM AMPACITY FOR COPPER AND ALUMINUM WIRE

Wire Gauge	Copper	Aluminum
14	25	0
12	30	25
10	40	35
8	55	45
6	75	60
4	95	75
2	130	100
1	150	115
1/0	170	135
2/0	265	150
4/0	360	205

Experienced robot designers will take this standard wire data chart with a grain of salt. First, there are substantial safety factors built into the current carrying capacities listed above. Also, for short lengths of wire, you will find that the current carrying capacity will be higher than for longer wire runs. Since the wiring runs on robots tend to be short, you may be able to use thinner gauges, and therefore lighter weight wire, and still carry a given amperage.

FACTORS OF SAFETY AND COMPETITIVE SITUATIONS

In the design of most products, engineers will include a factor of safety related to material properties to reduce in-the-field problems, but this often comes at the expense of added weight. In NASCAR competitions where those types of concerns are less important than getting the most horsepower and staying within the weight limits, builders often tend to design right up to the material limits of their components. Which type of robot builder are you?

6.
HOW TO CHOOSE A
FIGHTING ROBOT BATTERY

A basic fighting robot battery converts chemical energy into electric energy through an electrochemical process. Batteries are made from base units called "cells." A battery manufacturer can make a wide variety of voltages and capacities by combining cells in series or parallel combinations to create the desired voltage and output capacity. Also, there are several different kinds of batteries based on the elements and chemicals that make up the cell materials. The two most common types used in robot fighting are nickel cadmium, usually called NiCad, and sealed lead acid, often abbreviated SLA. NiCads have

Wild Child COURTESY TEAM WEASLE

6.1 Sealed lead acid batteries.

COURTESY OF HEPI

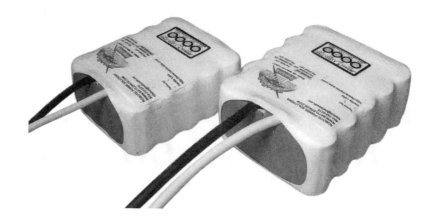

6.2 NiCad battery pack.

the advantage of being long-lasting and have the ability to discharge lots of energy in a hurry, if conditions warrant. SLAs are considerably less expensive and are available in many different capacities and voltage sizes.

Some builders use nickel metal hydride (NiMH) batteries in their robots. NiMH batteries have higher capacity ratings for a given size and weight than NiCads. However, they don't have the ability to source current (that is, discharge lots of energy very quickly) like NiCads.

The highest quality SLA batteries can source very high currents, but they have a lower capacity for a given weight versus NiCads. On the other hand, SLA batteries have no "memory" problems and can be charged faster. Memory problems occur when a NiCad battery is not fully discharged between charging cycles. This causes "memory effects," and the battery's permanent storage capacity becomes smaller.

Voltage, Load Current, and Amp-Hour Rating

Every DC motor is designed to operate at a specific voltage. Some small hobby motors happily whirl at top speed when connected to a single 1.5-volt alkaline cell battery. Other industrial-sized equipment motors require 100 volts or more to approach their design speeds. For fighting robots, the most common working motor voltages are 48, 36, 24, or 12 volts. As noted earlier, many bot builders intentionally overvoltage the motors in order to increase the horsepower output.

When choosing a battery, the first thing to know is what voltage you'll be using to power the motors. This is usually based on the voltage engraved on the chosen motor's identification nameplate, plus whatever overvoltage factor the builder is willing to risk.

The next thing to consider is the battery's energy storage capacity. The capacity of a battery is called its ampere-hour or amp-hour rating. For a given voltage, the bigger the amp-hour rating, the more total electrical energy it contains. The final item you want is the ability to get a lot of power out of the battery in a hurry, when you need it most. This "speed" is a consequence of the battery's internal resistance. NiCad batteries, for instance, typically have very low internal resistance characteristics, so they can disgorge all sorts of energy in a hurry if the need arises. For SLA batteries, some have much lower internal resistances than others, so this specification should be examined carefully by the robot designer before deciding on a particular battery. Internal resistance should be a key decision parameter when you choose your battery.

Because batteries tend to get hot during use, some robot builders go to the trouble of inserting small fans to blow cooling air on them during a contest. This helps reduce the battery's temperature and allows it to last longer. However, this may not be effective since it takes a lot of air to cool a hot battery. A better solution is to design the battery holder in such a way that it acts as a heat sink and wicks the excess heat away into the robot's frame. Whether fans or heat sinks are used, batteries must be held securely to the robot's frame. A battery holder is often made from tough metal, or a shop clamp can be used.

Some builders use long ratcheting plastic connectors called "tie-wraps" to attach batteries to the robot frame. While this is a common practice, I don't really recommend it because the heat built up could melt the tie-wraps. Similarly, special consideration should be paid to the heat built up on the batteries' screw or clip terminals.

In summary, NiCad batteries are lighter in weight than SLA

batteries and have the ability to source current very quickly. SLA batteries are less expensive than NiCads, but the specifications must be examined to make certain that a particular battery can source current quickly.

Determining the Right Battery for Your Robot

The voltage your robot needs depends primarily on the motors you're going to use. If you have a 12-volt motor, then you need a 12-volt battery (overvoltaging notwithstanding). And you want to look for motors with the smallest internal resistance so you can get the energy out of the batteries in a hurry if necessary.

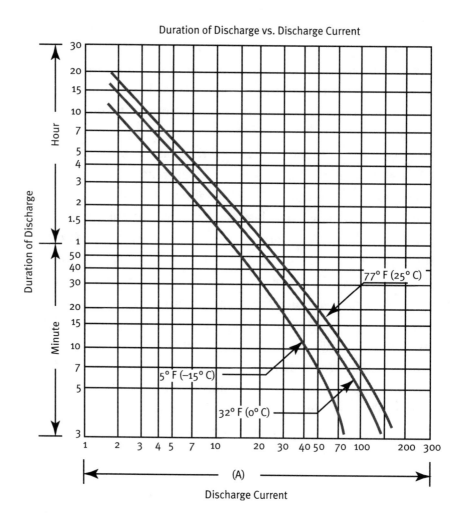

6.3 Generic 33 amp-hour battery specification sheet

The last item you need to specify is the amp-hour rating, and it's also the most difficult to determine. The bigger the amp-hour rating, the more energy you've got to use up during a match. Normally, bigger is better, except for one thing—batteries are very heavy! Can you afford to give up a great deal of weight toward your weight class maximum for excess battery weight? Most designers would rather put on more armor or bigger weapons than lug around excess batteries. The best thing to do is to determine how much battery you really need and use the right amount, with a reasonable safety factor.

This consideration is somewhat complex. One of the interesting characteristics of SLA batteries is that a battery disgorging a lot of power in a hurry is able to deliver less total energy than a battery

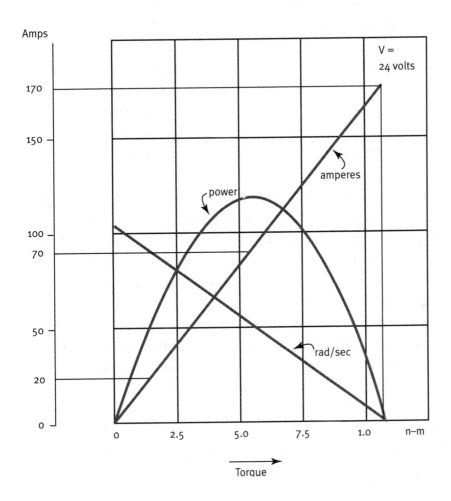

6.4 Typical ¾-horsepower DC motor data sheet

discharging its power more slowly. So you have to make allowances in total battery storage capacity if you think that the battery will be discharging energy in really big chunks for a substantial period of time. What you're going to do is to "de-rate" the battery as a function of its load current and really figure out what you need.

This can best be explained through an example. Let's say you're considering using a generic store-brand battery, a 12-volt, 33 amp-hour SLA battery to power a pair of $3/4$-horsepower DC permanent magnet motors. The specification sheet for the battery and motor is shown in Diagrams 6.3 and 6.4.

Assume that at stall torque, the motor's armature is locked up and, therefore, at the point of the maximum current drain. From the motor specification graph, we see the maximum amp draw is about 170 amps. From here, we'll make some reasonable assumptions about the current draw during the other battle phases described above and develop the following table:

Battle Phase	Amperage Drawn (from motor graph)
Stall	170 amps
Pushing	100 amps
Accelerating	70 amps
Constant Velocity	20 amps

From the battery data sheet for the 33 amp-hour battery, we see that if we discharged the battery at a constant 20 amps we'd expect to get almost 1 hour of battery life. From the same graph we see that if we discharged it at a constant 70 amps, this battery would last only 10 minutes. Therefore, the amp-hour rating of the battery would now be $1/6$ hour multiplied by 70 amps, or 12 amp-hours. Putting it another way, the big 33 amp-hour battery we thought we had is really only a 12 amp-hour box if we draw from it at a constant rate of 70 amps. That's a big loss in storage capacity! And it gets even worse at pushing and stall torques.

In order to determine the actual or "nominal" battery size you need to buy from the battery vendor, you need to de-rate the battery for each different battle phase. For example, if you assume that you consistently pulled 100 amps of juice out of the battery during the pushing phase, then you'd have to de-rate the battery by the percentage of time it spends in that phase, and so on for the other phases.

So how do you choose the correct size battery? By making a chart that describes the time spent in each battle phase that you expect to see in your style of fighting. You do this by making a chart like that shown on the following page. The second column shows four different robot activities and their expected current draw levels. Assign the percentage of time you expect your robot to be operating in each level during a match. The sum total of the numbers in the column at the far right provides the required amp-hour rating for the robot.

The table's fifth column provides the de-rating factor. It is determined by simply dividing the amount of time the robot spends in a phase by the amount of time the battery could support this phase. In the final column, the de-rating factor is multiplied by the nominal battery amp-hour rating to determine the amount of the battery's total nominal energy consumed during this phase. The sixth column provides the total amount of energy consumed.

If the energy consumed is smaller than the battery's nominal amp-hour rating, the battery is adequate. If the energy requirements are larger than the battery's storage capacity, then the battery is just too small for the job.

From the Bosch 750 motor and the generic battery specification graphs, you can put together the table on the following page.

The total energy consumed by each motor during a typical match is about 1.4 + 2.1 + 3.3 + 1.98 = 8.78 amp-hours. If the robot has two motors, then the battery will have to supply 2 x 8.78 amp-hours: 17.56 amp-hours. In this example, the 33 amp-hour battery is larger than required, so the robot designer may choose to use a smaller battery in terms of amp-hours, or keep the large battery in order to have a larger safety factor should the contest turn into a pushing match.

Another reason that NiCads are so popular among robot builders is that they don't have the same de-rating problems as SLA batteries. So, for a stated amp-hour rating, a NiCad will provide more minutes of use under high-discharge conditions than SLAs.

Battle Phase (assume a 180-second contest)	Number of minutes the robot spends engaged in this phase; and the percentage of the match spent in this activity level during 3-minute match	Current draw during this phase	Battery amp-minute rating at this level of current draw (from manufacturer's rating sheet)	De-Rating Factor = Time spent in phase/amp-hour rating at this level of current draw	Amount of equivalent energy capacity consumed during phase
Phase 1: Idling, moving at constant speed	1.67 minutes; 56% of the total match (based on experience; it could be higher or lower)	20 amps	40 minutes (obtained by looking up 20 amps on the manufacturer's rating sheet)	1.67 minutes / 40 minutes = 4.2%	4.2% x 33 amp-hours = 1.4 amp-hours
Phase 2: Accelerating, moving to position quickly	0.67 minutes; 22% of the total match	70 amps	11 minutes	0.67 minutes / 11 minutes = 6.1%	6.1% x 33 amp-hours = 2.1 amp-hours
Level 3: Pushing—pushing against another robot	0.50 minutes; 17% of the total match	120 amps	5 minutes	0.50 minutes or 17% of the total match	10% x 33 amp-hours = 3.3 amp-hours
Level 4: Stalled—maximum pushing, no moving	0.12 minutes; 4% of the total match	170 amps	2 minutes	0.12 minutes / 2 minutes = 6%	6% x 33 amp-hours = 1.98 amp-hours
				Total:	8.78 amp-hours

Connecting Batteries in Series and Parallel

If 24-volt motors are used, then two 12-volt batteries may be series connected in order to obtain the needed voltage. If 36 volts are desired, then three 12-volt batteries may be wired together in series, and so on. Robots builders may also wire batteries together in parallel in order to boost amp-hour ratings of the batteries. For example, a quantity of two 12-volt, 7.2 amp-hour batteries wired in parallel is equivalent to one 12-volt, 14.4 amp-hour battery.

Cutoff Switch

Although this was mentioned in the safety chapter, it is worth noting again that your robot must have a master cutoff switch or removable link that will immediately disconnect power from the batteries to all weapon systems and drive motors. The switch must be located such that it is easily turned on and off without putting the operator in danger while doing so.

Any switch accessible from the robot's exterior would work (theoretically), but there are some practical considerations that make this a slightly more complex matter than simply wiring in a knife switch.

First, the cutoff switch must be rated to handle the amperage supplied, even at stall torque. As discussed earlier, amperages can reach into the hundreds, or perhaps beyond, and the switch must be rated accordingly.

Second, the cutoff switch should be very dependable. It must not accidentally turn off during matches or turn on in the pit area. Some competitions require that a removable link be used to disconnect all power to the robot. A removable link and a safety cutoff switch are shown in Chapter 1.

Remember, the cutoff switch must cut off main battery power, not the control signal from the electronic speed controller or the relays. When the main disconnect is off, the robot is off, regardless of short circuits, mechanical damage to contacts, or anything else.

7.
MOTOR SPEED CONTROLLERS

It may be good to take a step back and look at how all the robot components fit together. Let's say you're at a robot competition with your new robot. Now at ringside, and holding your radio transmitter, it's time to start your robot moving forward and engage your competition. Let's examine this process closely.

7.1 Motor speed control block diagram

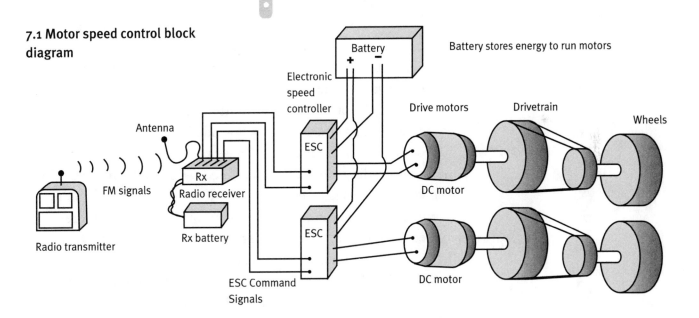

Battery stores energy to run motors

Radio: FM, FCM, or digital radio from R/C store or robotics supplier

Electronic speed controller: Device that controls current to motors, based on commands from ESC

Motors: DC electric

Drivetrain: Converts motor torque to linear motion

Sparafucile COURTESY TEAM
BACKYARD BALLISTICS

Diagram 7.1 shows the flow of information, electricity, and
power from your brain to the wheels of your robot. In sequence, it
works like this. Your brain tells your fingers to move the radio
transmitter joystick upward—the command for going forward. As
discussed in Chapter 4, the transmitter sends out a radio
command, in this case an FM radio-coded string of command
pulses of different frequencies. These radio waves are collected off
the robot's radio antenna and amplified by the radio receiver. The
receiver amplifies and separates the pulses into different command
channels and then sends the pulses out on two wires that are
connected to the motor controller. The motor controller looks at
the pulses and, based on their width, makes some decisions. It uses
its internal logic to direct current from the batteries to the motors.
The motor controller is basically a device that takes the tiny
voltages available from the radio receiver and controls large, even
huge, amperages, all based on the information contained in the tiny
current signals. If we used very, very small motors, like a radio-
controlled airplane servo motor, we wouldn't need a motor
controller, because the small amperage and voltage in the receiver

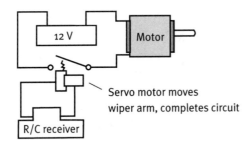

Servo motor moves
wiper arm, completes circuit

R/C receiver

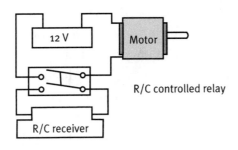

R/C controlled relay

R/C receiver

battery is large enough to drive the little servo. But motors used in all combat robots larger than a pound or two are too big and powerful and require power of greater magnitude than the receiver battery can provide.

The key thing to know, then, is that a robotic motor controller takes the information from the radio receiver and sends varying amounts of power from the batteries to one or all of the robot's motors. This chapter focuses on how a motor controller does its job. In real-life robot building, the motor speed controllers may well be the most complex and expensive piece of gear on your bot. They have a big job to do. They have to precisely control lots of amperage to each motor in ways that allow the driver to precisely control the robot's speed and direction.

Describing how a controller works sounds complicated, but if you dig into it a little, it's pretty straightforward. Take a look at the very simple control circuit shown in Diagram 7.3. In order to understand the diagram, you need to know what a relay is. A relay is an electrical component that mates the action of a big switch or contact closure with a little switch or contact closure. In other words, when the little switch is closed or opened, the relay closes or opens the big switch just the same.

In the diagram, the bot driver pulls the stick up to transmit a "go forward" command on the radio. The transmitter sends the coded pulses that are "heard" by the radio receiver, and this sends the commands to a radio-control interface relay. The internal logic within this type of relay sees the pulses and closes a relay control. When that closes, it also closes the circuit between the big batteries and the big robot motors. This done, the robot goes forward.

At this point, you have a control system for a robot that allows it to go forward when the relay is closed, and stops it from going forward when the relay is opened. Obviously, we want our robot to go forward, backward, left, and right, and move at various speeds. All of these needs make the robot control scheme more complicated, yet each is just a further extension of the original idea. Diagram 7.4 shows how to use two R/C relays to make your robot go forward or backward.

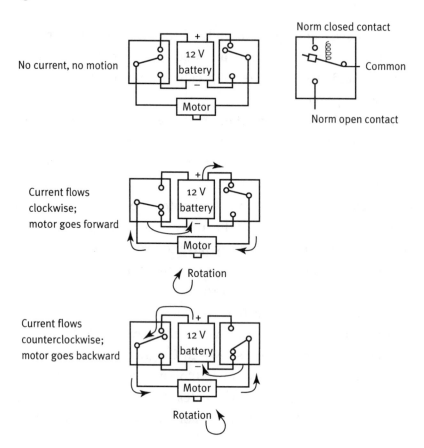

7.4 Forward and reverse directional control

Steering Systems

There are two different types of drive systems that can be used to steer your robot. The first and most common method is called "skid steer," where one motor drives each wheel in a two-wheel drive system. To turn the robot, a different amount of power is provided to each wheel. If the right has more power, the robot turns left. If the left has more power, the robot turns right. Most builders opt for skid steer as it is both easier to design and build than "car steer."

In car steer, one motor turns both of the drive wheels. At the same time, a separate steering mechanism controls the direction of an axle. Unlike a Ford pickup truck, a motor is used to control direction of the axle instead of a steering wheel manipulated by human hands. Car steer is considerably more complex to implement than skid steer. If the steering wheels are also the driving wheels, then a speed compensating mechanism called a differential must be used. Differentials are complicated, generally well beyond the capability of most builders, so they are not covered here.

Going Backward, Slowly

The relay control diagram, Diagram 7.5, shows how to use four relays to make your robot go forward, left, right, and reverse using skid steering. This circuit, using four relays, is called an H-bridge circuit and is very commonly used to control robots.

The H-bridge does what every robot builder needs—it allows you to drive forward, backward, left, and right. Oftentimes, resourceful robot builders use automotive relays, such as Ford starter motor relays. These are available at automotive parts stores. Automotive relays can handle more than 500 amps, although they're not really designed to handle such current for long periods of time. A full 200-amp capacity H-bridge can cost as little as $50, but up to $100.

So a relay-type H-bridge using automotive starter relays is the way to go, right? Well, not necessarily. There are two issues with a

7.5 H-Bridge control of two motors

H-Bridge Control circuit independently controls two motors. Robot turns counterclockwise with relays in position shown.

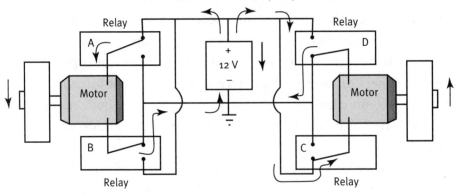

Robot moves clockwise with relays in position shown.

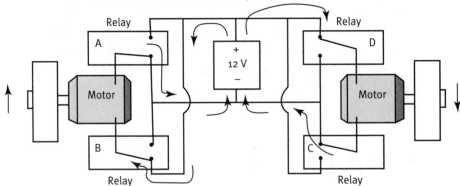

relay-based speed control solution. The first issue is that an automotive relay user will have to buy or design an interface circuit that will allow the driver to control the relay from the R/C receiver on board the robot.

The second issue is that many builders also want to go faster and slower depending on the situation in the arena during the contest. Relay-type H-bridges by themselves can't do that. They are either completely open or completely closed. Relays are black-or-white kinds of devices—they don't work in shades of gray.

In order to vary the speed of the robot, we need something called "proportional speed control." Proportional speed control allows the driver to send a little power through to the wheels—a

A top heavyweight lifter: Toro. COURTESY BATTLEBOTS, INC., AND INERTIA LABS

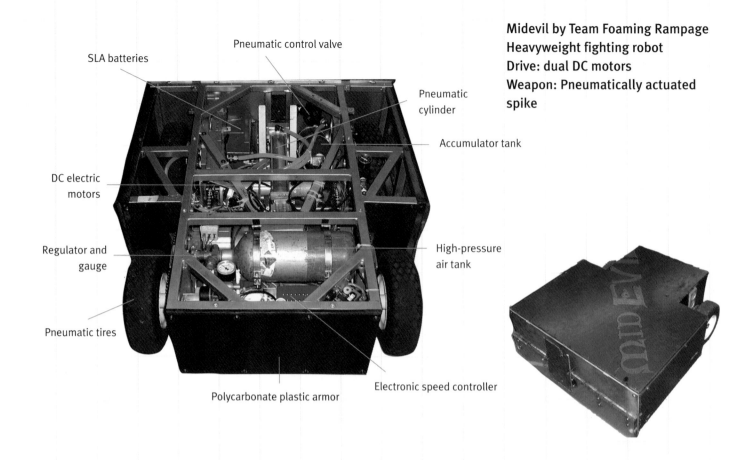

SLA batteries

Pneumatic control valve

Midevil by Team Foaming Rampage
Heavyweight fighting robot
Drive: dual DC motors
Weapon: Pneumatically actuated
spike

Pneumatic cylinder

Accumulator tank

DC electric motors

Regulator and gauge

High-pressure air tank

Pneumatic tires

Polycarbonate plastic armor

Electronic speed controller

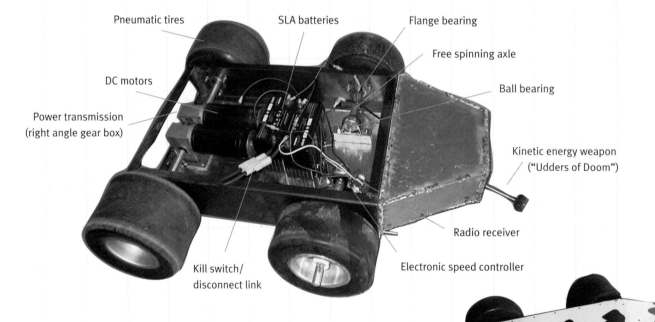

Pneumatic tires

SLA batteries

Flange bearing

Free spinning axle

DC motors

Ball bearing

Power transmission (right angle gear box)

Kinetic energy weapon ("Udders of Doom")

Kill switch/ disconnect link

Radio receiver

Electronic speed controller

Mad Cow by Team Rabid Robotics
Middleweight robot
Drive: Dual DC permanent magnet
motors with built-in right angle gear box
Weapon: Kinetic energy "Udders of Doom"

Battle-hardened, kinetic energy champion: Dr. Inferno Jr. COURTESY BATTLEBOTS, INC., AND JASON DANTE BARDIS

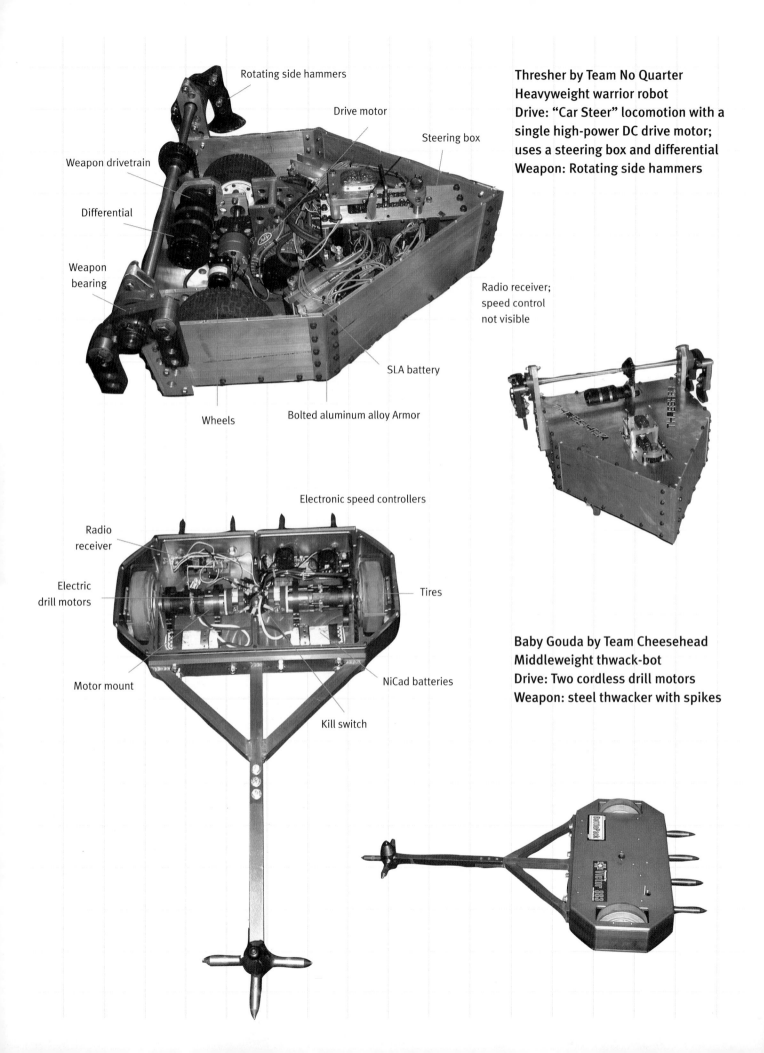

Rotating side hammers

Drive motor

Steering box

Weapon drivetrain

Differential

Weapon
bearing

Thresher by Team No Quarter
Heavyweight warrior robot
Drive: "Car Steer" locomotion with a
single high-power DC drive motor;
uses a steering box and differential
Weapon: Rotating side hammers

Radio receiver;
speed control
not visible

SLA battery

Wheels

Bolted aluminum alloy Armor

Electronic speed controllers

Radio
receiver

Electric
drill motors

Tires

Motor mount

NiCad batteries

Kill switch

Baby Gouda by Team Cheesehead
Middleweight thwack-bot
Drive: Two cordless drill motors
Weapon: steel thwacker with spikes

Good Robots Built on a Budget

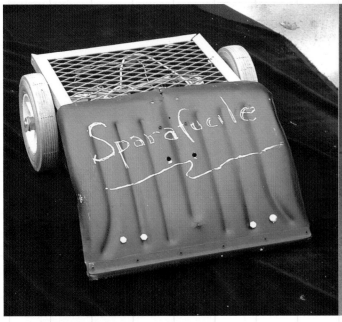

Sparafucile, a BMW. COURTESY TEAM BACKYARD BALLISTICS

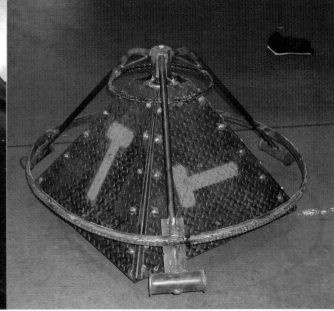

Thor, a spinner. COURTESY TEAM NORSE

Ricky Ticky, a BMW. COURTESY TEAM NEVERMIND

Al, a basic BMW. COURTESY TEAM BOBBING FOR FRENCH FRIES

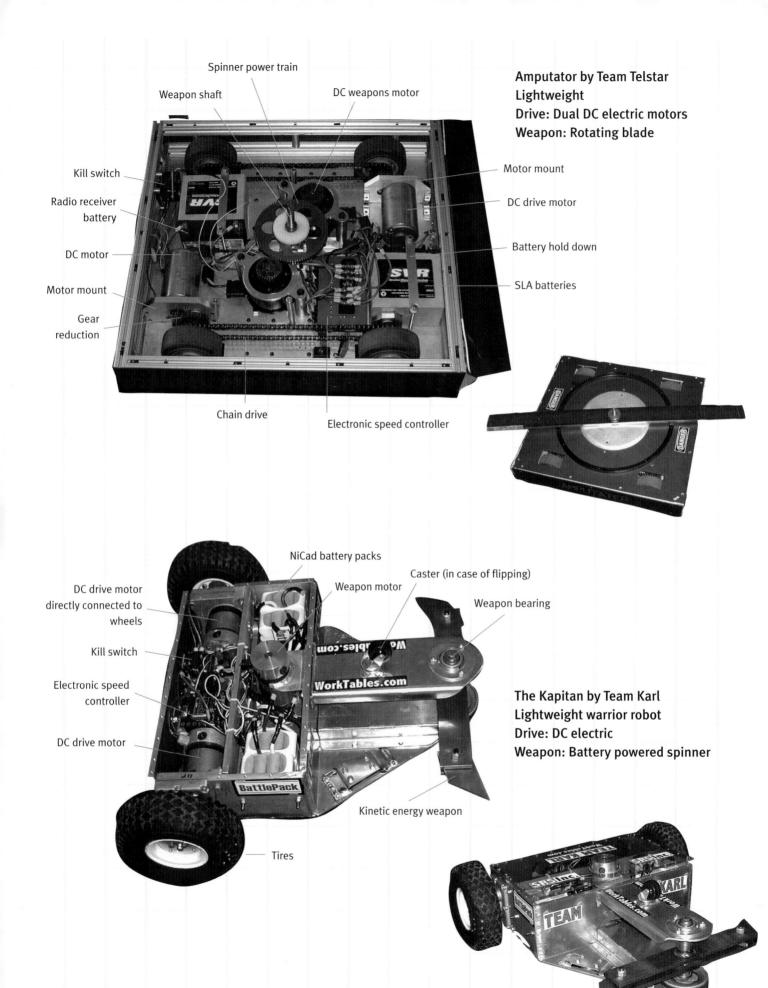

Spinner power train

Weapon shaft

DC weapons motor

Kill switch

Radio receiver battery

DC motor

Motor mount

Gear reduction

Amputator by Team Telstar
Lightweight
Drive: Dual DC electric motors
Weapon: Rotating blade

Motor mount

DC drive motor

Battery hold down

SLA batteries

Chain drive

Electronic speed controller

NiCad battery packs

Weapon motor

Caster (in case of flipping)

Weapon bearing

DC drive motor directly connected to wheels

Kill switch

Electronic speed controller

DC drive motor

The Kapitan by Team Karl
Lightweight warrior robot
Drive: DC electric
Weapon: Battery powered spinner

Kinetic energy weapon

Tires

Well-designed destruction from a hybrid thwack-bot: Tazbot. COURTESY BATTLEBOTS, INC., AND DONALD HUTSON

Stinger, a BMW. COURTESY A. J. KLEIN OSOWSKI

A lightweight, wedge-shaped BMW: Wedge Wood.

COURTESY TEAM FUBAR

Peapod, a thwack-bot. COURTESY TEAM NEVERMIND

Chameleon bot with attachments: Psychotic Reaction.

COURTESY TEAM KONTROLLED KAOS

medium amount, or all of it. Lift the joystick up a little, and you go forward slowly. Lift the joystick up all the way, and you go at maximum speed. Relays won't allow this, so you need to put something in front of your H-bridge that meters out the current based on how far the control stick is pulled. Most controllers accomplish this using a concept called pulse width modulation (PWM).

PROPORTIONAL SPEED CONTROLLERS VERSUS RELAY CONTROLLERS

Attribute	Differences
Degree of speed control	Proportional speed controllers provide an infinitely variable range of speeds, so the robot can go slow, fast, or any speed in between. The relays are on/off devices, so either your robot receives full power or no power at all.
Joystick	Some proportional speed controllers, like the Vantec models, allow single-stick on your R/C controller for all directions right out of the box. The relays require a separate stick for the left-side and the right-side motors. This is not terrible, but it takes more skill to drive, and makes it harder to control the weapon simultaneously.
Cost	The proportional speed controller is four to eight times the cost of a relay-based solution.

Pulse Width Modulation

Pulse width modulation, or PWM, is a common method of adding degrees of gray or proportion in on/off environments. To understand PWM, consider the spinning rod motor controller shown in the mechanical speed controller diagram below. It is made out of an assembly of conducting copper and nonconducting plastic and is rotated at high speed by a very small motor. Brushes have been put on either side of the spinning rod, just like the brushes on a DC motor. The spinning rod also moves back and forth along its longitudinal axis according to the position of a second motor, this time a servo motor.

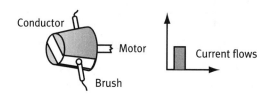

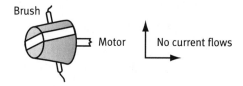

As the truncated cone spins, the brushes make and lose contact with the electrical conductor. Moving the brushes left increases the proportion of conducting time and provides a higher average voltage at the brush leads.

7.6 Mechanical speed controller

As the motor spins, the brushes alternately make an electrical connection when they contact the copper and then lose contact when they contact the plastic. When the brush-to-brush connection is all copper, all of the current in the batteries is directed through the copper to the motors. When the brush-to-brush connection is electrically blocked by the plastic, no current flows and no power is derived.

Because the tube-spinning motor spins the shaft so quickly, the amount of power delivered to the wheel-driving motors is blurred together from the wheel's perspective, and is just the average of the on time divided by the off time. *Average* is the key word here, because that is how proportional control is obtained using on/off devices.

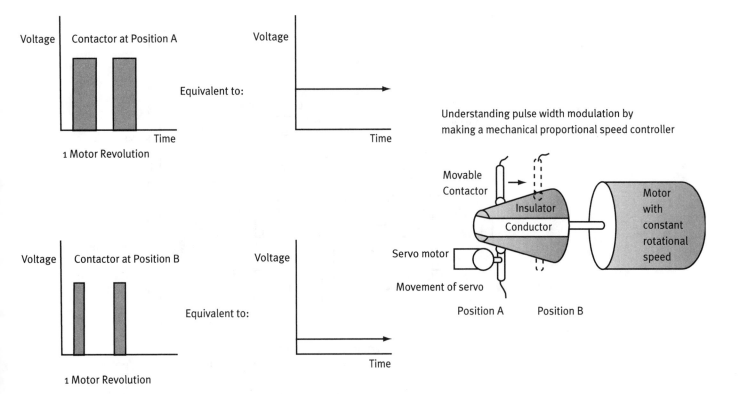

Voltage | Contactor at Position A

1 Motor Revolution

Voltage

Equivalent to:

Time

Voltage | Contactor at Position B

1 Motor Revolution

Voltage

Equivalent to:

Time

Understanding pulse width modulation by making a mechanical proportional speed controller

Movable Contactor

Insulator

Conductor

Servo motor

Movement of servo

Position A   Position B

Motor with constant rotational speed

Explanation: When the cone rotates, the contacting brushes make and then lose electrical contact. The servo pushes the brush out to a position where the average contact time is lower. This lowers the average voltage and decreases motor speed.

7.7 High and low power examples

The servo motor pushes the tube in and out. Because of the V-shaped interface between copper and plastic, the amount of time the moving brush is in contact with the conducting copper is controllable by the servo motor positioning. When the servo is in the fully "in" position (Position A in Diagram 7.7), the average time that electricity is conducted is high, resulting in high current flows. When the servo is in the fully "out" position (Position B), the average contact time is low, resulting in low average current flows. This type of motor control, using the average of rapidly repeating on and off pulses of current, is called pulse width modulation, or PWM.

Diagram 7.8 shows an electromechanical proportional motor controller that can make a robot go forward, backward, left, and right, at any speed between full stop and maximum.

Commercial speed controllers work in a similar fashion. Instead of a rotating copper/plastic tube, they use a grouping of transistors

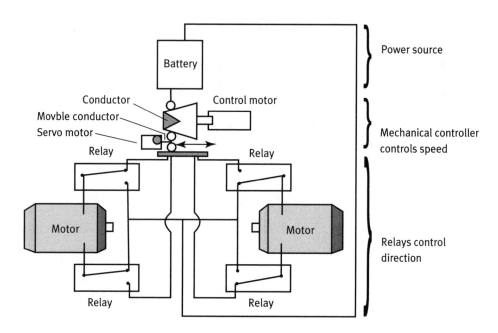

Battery

Power source

Conductor
Movble conductor
Servo motor
Control motor

Relay   Relay

Mechanical controller
controls speed

Motor   Motor

Relay   Relay

Relays control
direction

Directional and proportional speed
control of a robot using mechanical parts.

7.8 Mechanical proportional speed controller with H-bridge

on integrated circuits called MOSFETs that are capable of switching big current loads on and off at high speed. There are no whirling connectors or brushes in an integrated circuit-based speed controller. The solid-state circuitry uses various integrated circuit components to obtain PWM. But in the end, the result is the same. This drawing shows a schematic diagram for a transistor-based electronic speed controller.

In this case, the large, motor-driving current flows from the source to the drain of the MOSFET depending upon a much smaller current at the transistor gate. (The source is the point on the transistor where the large, controlled current enters. The drain is where it leaves, and the gate is the point where the smaller, controlling current enters the transistor.) Some high-end radios and some commercial speed controllers, such as those made by Vantec (see Appendix I), have the added advantage of providing circuitry that includes a mixing function. Mixing will allow a single joystick to control the robot's direction by "mixing" the action of two individual motor controllers into a single control lever.

Designing a motor speed controller using integrated circuits is a difficult engineering endeavor; a small number of builders who are

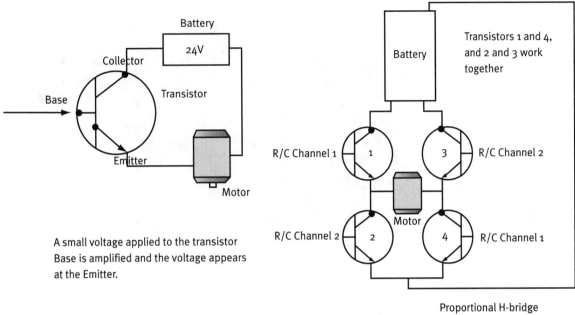

A small voltage applied to the transistor
Base is amplified and the voltage appears
at the Emitter.

Transistors 1 and 4,
and 2 and 3 work
together

Proportional H-bridge
motor control circuit

**7.9 Transistor-based
proportional speed controller**

adept in electronic hardware design do, however, make their own
speed controllers. See the ESC section in Appendix I for information
on obtaining pre-made circuit boards if you're interested in making
your own. For precise control of electric motors, most people have
to purchase a commercially available electronic speed controller.
There are several on the market from Vantec, 4QD, and Innovation
First. Speed controllers are not difficult to wire into your robot.
Most manufacturers provide reasonably good instructions for their
products. Be aware that speed controllers are expensive and do not
like being connected in circuits incorrectly. Reversing polarity may
destroy your speed controller, so always be careful in making
connections. Some speed controllers, like those from 4QD and
Innovation First, will require an interface board to decode the
signal from the radio receiver. (See Appendix I for information on
obtaining speed controllers and interfaces.)

Before you run out and order a speed controller for your robot,
you will need to know a few things. First, what voltage do your
motors require? Unlike the wink-wink business that occurs when
builders apply higher than rated voltages to many DC motors,
overvoltaging a solid-state motor controller is a very bad idea. It

7.10 Speed controller. COURTESY OF VANTEC CORP.

generally results in something burning up, costing you a good deal of frustration and money.

Second, what is the motor's current draw at stall condition? Speed controllers are specified to handle a certain amount of current and no more. You could simply look at your motor's stall current draw and choose a motor controller that is rated at that level. However—and this is an important "however"—many motors draw incredible amounts of current at stall torque, as much as 400 percent of the continuous normal maximum level. Obtaining a solid state, integrated circuit power controller capable of switching, say, 300 amps would cost a great deal of money.

But it may be that you don't need a speed controller capable of switching your motor's stall current. Many robots are designed in such a way that the motor will never reach stall torque. Even bumping into a brick wall won't stall out the motors because the wheels slip and continue to spin, or the drive chain breaks, or something else happens to prevent the motor armature from actually locking up. If motor stall never occurs, then you don't have to design the motor controller to handle the stall current. Instead, it can be something less. How much less? Well, it depends. One approach is to determine your motor's continuous maximum current rating, which is often provided on the motor data sheet. Or the builder can make some reasoned estimates regarding a percentage of how close the motor is going to get to stalling and then pick the appropriate maximum amperage based on the torque-current draw graph on the motor data sheet.

Another design method is to procure a speed controller with integral current-limiting circuitry. In this scheme, the amount of current switched by the controller is self-limited to a safe (for the controller) maximum. The downside of self-limiting, of course, is that the motors will not get an opportunity to produce their maximum horsepower rating, even for a short period of time. Practical engineering is an art as well as a science, and choosing a motor speed controller is another example of that fact.

TC MECHWARS III

St. Paul, Minnesota

Jon VenderVelde wants to be a stand-up comedian, but today he's about as far away from mirthful as one can get without being locked in a Turkish prison. The beat-up white trailer truck that stores the polycarbonate and steel arena for the fighting robot tournament that he promotes has just gone missing. Missing? Yup, a person or persons unknown swiped the truck, and now it's gone. Even worse, the third annual Mechwars tournament is scheduled to start in just three days.

Maybe it wasn't really stolen, but merely towed somewhere for illegal parking. A series of frantic calls to every towing company in the area yields no fruit. Stolen it must be, but how hard can it be to find an old truck towing a 24-foot-long trailer full of robot parts?

Pretty hard, evidently, as the cops have no idea where it is, but they tell Jon they'll call just as soon as it's spotted. This makes for some hard choices. Big money has been laid out to rent the auditorium space. People are already boarding airplanes and attaching small trailers to their cars in order to head to Minneapolis for Mechwars.

The tournament deathwatch begins. Some builders volunteer to drive the backstreets of Minneapolis hoping to spy the wayward truck. Time passes, more calls are made to more towing companies and more police departments, but nothing turns up. The deadline passes. The show is cancelled.

Jon is resilient, and, just after the cancellation deadline, the missing truck turns up in a seedy neighborhood, its cargo more or less intact. Jon reschedules, hoping the calamity hasn't damaged his credibility with builders and audience beyond redemption.

January 19 is a much better day for the Mechwars staff. Not only has the truck stayed found, but the new location, the National Guard Armory in St. Paul, is a big step up from the original venue.

This will be the third Mechwars tournament. The second annual Mechwars was successful, but did have several, well, kinks. The weight limits for superheavyweight bots at TC Mechwars top out at 370 pounds, more than 30 pounds more than the highest weight limit for wheeled robots at the national tournaments. A 370-pound robot, bristling with iron spikes, is scary in a dead run. It's strong enough to crash through a house, which made the margin of safety on the plastic arena, the one that went AWOL in October, way too thin for my liking. More than once in Mechwars II, a robot crashed through the arena walls, coming close to the spectators. Everything would then stop for 20 minutes while the stage crew took to repairing the arena, mostly with pole spikes and sledgehammers. While moderately interesting at first, watching John Henry repeatedly swing 12 pounds of steel in frantic efforts to quickly bang the arena back together wearies the crowd. Although Vandervelde and his assistants built a new, stronger design for the Mechwars III arena, the smart folks are still sitting in the back row for this one.

As at Battlebots, each competitor's robot is weighed and inspected for adherence to the published safety rules. Mechwars devises a lot of its own tournament procedures by modifying rules from the R/C airplane races, paintball tournaments, and model rocket meets. For instance, each competitor requests a specific frequency on which to transmit. Since there are only a small number of frequencies available, there are multiple competitors operating on the same frequency band. To partially solve this problem, frequency tags were handed out in early tournaments, and only builders in possession of a tag were to have an operating transmitter.

Mechwars, like other small regional tournaments, improves each time it is held. The third time around it is handled quite professionally, and nationally known robots such New Cruelty, Tortoise, and Son of Whyachi, are present to wow the crowd. Son of Whyachi is widely known as a "crazy powerful" spinning robot, so much so that one builder compares the aftermath of its hammer blows to what remains when a Hefty bag filled with vegetable soup is thrown off a six-story building.

The quality of the robots and the venue impresses the crowd and here, out in the robot-building hinterlands of the Midwest, the robots compare well with the best at the national events. Robotic combat tournaments are springing up like mushrooms in a cow pasture after a warm spring rain. Besides the nationally televised contests like *Robot Wars, Robotica,* and *Battlebots,* local events like Mechwars, SORC, and Robot Clash are becoming more and more popular.

The burgeoning popularity of the sport provides the robot-builder community myriad opportunities to try their latest creations, wherever they are. One question comes up often at tournaments. How long will this whole thing last? Will robot fighting go the way of riding mechanical bulls and playing pogs? The Mechwars participants, at least, think not. It is, they say, more substantial, and presents more of an intellectual challenge and more opportunity for self-expression. Some observers describe the warrior robot phenomenon as a cross between professional wrestling and NASCAR racing. Both of these activities have passed the test of time, and the builders and promoters of warrior robots hope for the same.

8.
GASOLINE ENGINES

Some warrior robots builders make very good use of gasoline engines. The internal combustion engine, or IC, is best suited for a very specific use—it makes an excellent prime mover for large kinetic energy weapons, such as rotary saw blades and lawnmower blade top spinners. If you compare the performance of DC electric motors with IC engines, you'll find each has some important advantages in special situations. Before you run out and dismantle the trusty Lawn Boy, here are some things to consider.

Advantages

1. Internal combustion engines generally have better power-to-weight ratios than electric motors. You can get more horsepower per pound from an IC engine. A standard, off-the-shelf IC engine is inexpensive compared to most multiple-horsepower DC electric motors of similar size.

2. IC engines don't require heavy SLA or NiCad battery packs. A few ounces of gasoline are more than enough to power the engine for the full length of a match. All in all, a 7-horsepower IC engine typically weighs about 35 pounds, so at least for the 5 to 7 horsepower ranges, you can expect about 5 horsepower per pound. Compare that with an electric motor. Sure, there are

some big electric motors that put out 8 horsepower or so and they only weigh 22 pounds, but these beasts need lots of batteries to run a full 3 minutes, which easily adds another 60 pounds into the equation. For an electric motor, 8 horsepower/82 pounds is about 0.1 hp/pound. But for an IC engine, 7 horsepower/35 pounds is about 0.2 hp/pound. In other words, gasoline motors pack about twice the punch per pound as electric motors.

3. Designing an IC engine speed controller is simple. All it takes is a good quality servo and a centrifugal clutch from a go-kart.

4. IC engines are louder and therefore may be more impressive to the judges and the audience.

Disadvantages

1. It's very difficult to design a drivetrain for an IC engine that has the ability to reverse the motor's direction. While many DC motors can reverse simply by reversing the direction of the current flow, an IC engine requires a complicated reversing power transmission. Since any motor that powers the main wheels must have the ability to go backward, IC engines are impractical for most fighting robot drive applications.

2. Engine speed for IC motors is more difficult to control than for DC motors.

3. Although exceptions are out there, IC motors are designed to work in an upright position. If the robot gets flipped in a competition, it will stop working as soon as the carburetor runs out of fuel.

4. It is difficult to design a way to restart a stalled motor during a match. An onboard starter adds weight and quite a bit of complexity. Without an R/C onboard starter system, once an IC engine quits, it's pretty much relegated to ballast status for the duration of the match.

So is there a place for internal combustion engines on your robot? Maybe. If your choice of weapon is a large spinning kinetic energy or cutting blade, and you are willing to forgo the ability to drive it after being flipped, an IC engine is a good choice.

Internal Combustion Engine Speed Control

In its simplest incarnation, the IC engine drivetrain consists of the IC engine with a vertical driveshaft and spring-loaded throttle mechanism (like the kind found on most lawnmowers and some snowblowers), a speed-governing system, a centrifugal clutch, and a mounted servo motor.

The throttle controls the amount of gasoline entering the engine intake by opening and closing a flap above the carburetor's venturi nozzle. The flaps are spring-loaded, so if the throttle link breaks, it automatically closes the carburetor flap and makes the engine run at its minimum, or idle, speed. The driver controls the position of the throttle by simply attaching the throttle link directly to the servo motor control wheel.

Using a servo motor to control the speed of an IC engine can be a fairly difficult endeavor, depending on the size of your engine and the size of the weapon it is turning. This is because a spinning, heavy weapon directly linked to a motor driveshaft can cause the motor to rev up, sometimes uncontrollably. The spinning weapon is really a flywheel, with a large moment of inertia. Due to the flywheel effect, the faster the blade turns, the faster the motor wants to rev.

To remedy that problem, heavy spinning weapons connected to gasoline engines need to have a way to control the speed on the engine. A mechanical speed control or speed governor may be necessary. Builders who choose to use IC engines often design a linkage composed of springs, levers, and mechanical attachments between the servo and the engine linkage. The design that works best depends on the particular motor and control components the builder uses.

Some builders place a spring in the connection between the servo motor and the throttle that makes the motor automatically accelerate at full power until it reaches a certain RPM, when the governor overcomes the pull of the spring and throttles-back the motor. During a robot fight, when RPM drops after a hit, the spring causes the motor to automatically return to full power until the design RPM is again achieved.

A safety concern related to the use of IC engines is that some tournaments have regulations that require the IC engine motor to go to idle, or even stop, if the control signal is lost. In this situation, the best way to accomplish this is to use a PCM radio or digital radio controller that has an onboard microprocessor. If the signal is lost, then the onboard controller can be programmed to instruct the servo controlling the throttle to shut down. For more stringent safety conditions, the PCM controller can kill the connectivity to the spark plug to shut down the motor completely if the signal is lost.

Clutches

Most competitions have rules that state that at the start of any match, all weapons must be completely stationary. Furthermore, the weapon-driving motor should fail safely. That is, if radio control to the robot is lost, the blade should stop spinning. This poses a problem. The throttle control should go to minimum speed in both situations because the throttle intake is at its most restricted. But even with the throttle flap completely closed the engine still turns the blade, albeit slowly. This is still an illegal setup. So robot builders with IC engines use a device called a centrifugal clutch.

The centrifugal clutch is the link between the engine driveshaft and the shaft that spins the weapon. The clutch's purpose is to disengage one shaft from the other when the engine is at a low-speed idle so that the saw blade or the chain does not move. The clutch disengages the shafts at low speeds. But when the driver moves the servo motor controller to the engaged position, the engine will rev up, because the servo motor pulls the carburetor flap open and the engine gets more gas. When this occurs, the centrifugal clutch engages and the weapon turns.

A typical centrifugal clutch consists of three main parts:

1. An outer clutch surface that turns freely. This surface is the inside of a drum that is either welded to the main spinning weapon or is connected to a chain sprocket that drives a chain that is in turn attached to the weapon. In either event, when the drum spins, the weapon spins.

2. A center shaft attached directly to the engine's driveshaft. If the engine is turning, so is the shaft.

3. This is the important part: there are a number of movable clutch weights attached to the center shaft, along with a spring that keeps them retracted against the shaft.

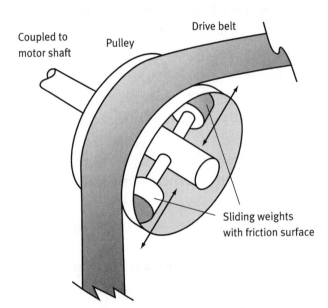

Coupled to motor shaft

Pulley

Drive belt

As the motor shaft turns faster, the sliding weights are pushed outboard by centrifugal force. The contact surfaces on the weights turn the pulley.

Sliding weights with friction surface

8.1 Centrifugal clutch

The center shaft and weights spin as one. If they are spinning slowly enough, the weights continue to hug the center shaft as they are held against it by the force of the spring. But when the engine spins fast enough, the centrifugal force on the weights (hence the name, centrifugal clutch) overcomes the inward force applied by the spring, and the weights move outward. As the driveshaft continues to spin faster, the moving weights come into contact with the inside of the drum, so the drum starts to spin. The drum, weights, and center shaft become a single spinning unit because of the friction between the weights and the drum. Once the drum starts turning, so does the weapon. Robot builders use centrifugal clutches because they are automatic. In a car with a manual transmission, you need a clutch pedal. A centrifugal clutch doesn't. It slips automatically to avoid stalling the engine when the engine slows down. Once the engine is spinning fast enough, most of the power is transmitted through the clutch from the motor to the weapon. If the spinning blade encounters an immovable object, with luck, the clutch will simply slip instead of stalling out the motor.

Mounting a Weapon to a Motor Shaft

It takes a certain amount of ingenuity to adequately mount a weapon to a spinning motor shaft. The connection between the motor shaft and items such as gears, sprockets, pulleys, and saw blades is subject to tremendous impact loads and stresses. This is the spot that most frequently fails on robots.

Tips for mounting weapons to shafts:

1. Avoid set screws. They are generally not strong enough to adequately address the shock loadings associated with robot combat. Instead, use a keyed shaft and a square key, or consider drilling a hole completely through the shaft and then inserting a bolt through the collar area.
2. One possibility for mounting a weapon is to weld the item directly to the shaft, if possible. Welding is permanent, so it may

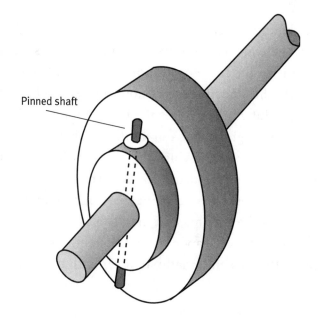

Pinned shaft

8.2 Attaching items to shaft using a pin

BUILDING BOTS

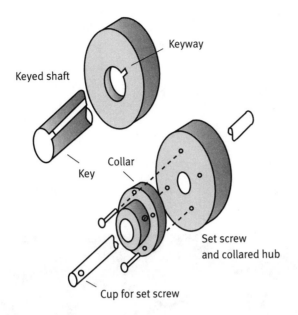

Keyed shaft

Keyway

Key

Collar

Set screw
and collared hub

Cup for set screw

be unsuitable if your overall design requires disassembly of the components from the driving shaft. Instead, many builders will use a shaft collar or a key to mount items to spinning shafts.

3. A hole may be drilled in the end of a shaft and then tapped for a bolt and washer. A metal-cutting lathe is generally required for cutting such holes, as it will be difficult to align a drill press with enough precision.

4. See Chapter 9 on drivetrains for examples on sizing shafts and other powertrain components.

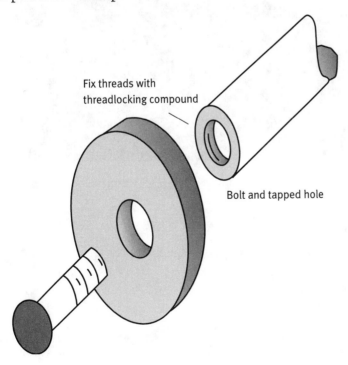

Fix threads with
threadlocking compound

Bolt and tapped hole

8.4 Attaching items to shaft
using a bolt and tapped hole

9.
DRIVETRAINS

A fighting robot drivetrain is a designed system of wheels, shafts, and speed- and torque-transmitting equipment that takes power from the robot's motor shafts and applies it to the driving wheels. If you look back at Chapter 5, you'll see how electrical energy is converted to the spinning motion of the motor rotor or armature. The speeds at which the armature of a DC motor turns are normally too high for the motor to be connected directly to the wheels. If a builder were to directly connect a motor to the drive wheels (not a gearhead motor, which includes a drivetrain within the motor housing), he or she would find that at such high speeds, the robot would have very little torque, acceleration, and controlability. Therefore, speed-reducing equipment is required to produce a usable combination of rotational speed and torque.

Setting Expectations

Designing a fighting robot drivetrain is not a cookbook endeavor. The builder must determine the right components based on the huge variety of motors, wheels, and weapons available. Plus the design of mechanical components in any drivetrain is very complicated. To do it like a real engineer would require delving into equations, theories of wear and performance, product data sheets

Mad Cow COURTESY TEAM RABID ROBOTICS

and catalogs, prototyping, and possibly buying a pocket protector. It takes years of study and experience to be a competent mechanical engineer, so don't expect a single chapter in a single book to provide anything more than a high-level introduction into designing this kind of stuff. Think of this section as more an "engineer's notebook" than a series of step-by-step instructions to build a drivetrain.

Overview

You can consider a drivetrain a conversion machine with inputs and outputs. The inputs are the rotational speed and torque of the motor shaft, and the outputs are the speed and torque of the wheels. They vary inversely, according to the size and design of the drivetrain components. There are many ways to convert speed and torque, but most robot builders utilize gears, pulleys, and belts, and, most frequently, roller chain and sprockets. Each type of speed and torque conversion method has its own advantages and disadvantages, and therefore its own body of adherents and detractors.

The torque- and speed-converting components need to be held in strict spatial relationships with one another. This is the job of the supporting components. Support components such as shafts, bearings, and couplings fix the motors, wheels, and conversion components in the desired dimensional relationships with one another. What supports the wheels? The shaft. What supports the shaft? The bearings. What supports the bearings? The frame, and so on. Let's begin discussing the primary supporting components: shafts and bearings.

Shafts and Bearings

A shaft is a (usually) solid metal cylinder designed to support wheels, gears, and sprockets. As the shaft turns, it also imparts a rotation to whatever is attached to it. So an important concern is how to fix a wheel or other drivetrain component to it without slipping. There are several ways to do this, the two most common being a key or a setscrew.

A key is a small bar of metal made to fit into a precisely machined groove on a shaft as shown in the drawing. Keyways are difficult to machine but are very solid and secure. In fact, some say a key is absolutely the best method for attaching something to a shaft. Cutting a keyway in a shaft requires a machine tool like a mill or a broach, but it is more common for builders to simply buy shafting with the keyway precut.

A setscrew is a small, headless screw that affixes a wheel or gear's collar to a shaft by friction. Setscrews use friction acting over a fairly small area—the threads—so they tend to slip, making them generally less desirable for high-torque situations. Some builders fix things like wheels to shafts by drilling a hole through the shaft and wheel and then inserting a pin or rod to hold everything in place. This is crude but cheap, and it works fine. The hole can be drilled either perpendicularly or horizontally, depending on how you want to insert the pin.

Alternatively, if the hole in a component is just slightly smaller than the shaft diameter, it is possible to press fit (whack with a hammer) one into the other and obtain a non-slipping fit. However, it is not easily disassembled, and the press fit may start to slip, at which point the fighting robot has big problems. How does a builder choose the right shaft? Since the torques in lightweight and middleweight robots do not usually approach design limits for most shafts, the easiest way to choose a shaft is to simply use a steel shaft of a diameter that mates with your wheels without needing bushings or adapters. But if you have a big, powerful robot, you can engineer it.

Determining How Much Horsepower Your Driveshaft Can Carry

In order to determine how much torque your driveshaft can carry, first determine how much shear stress it can handle. A commonly used rule of thumb for determining the maximum allowable stress on a shaft subject to high shock loading (as in a fighting robot environment) is to use a shear stress rating of 8,000 psi for steel shafts if no keyway is present, and 6,000 psi if there is a keyway. The following formula can be used to calculate the maximum torque you can put on a steel shaft of a given diameter:

$$T_{max} \text{ (inch-pounds)} = \frac{\text{Shear Stress Rating} \times \pi \times \text{Diameter (inches)}^3}{16 \times K_t}$$

Where d is diameter of shaft, K_t is a shock load de-rating factor (1 is no-shock, 3 is high-shock environment), and T_{max} is the torque applied during shock loading. Robot combat is a high-shock environment, so use the maximum de-rating factor of 3.

Let's use an example to see how we can determine the maximum horsepower our shaft can handle. Assume we're using a motor with:

- a ½-inch diameter non-alloy steel shaft, so $d^3 = (1/2)^3 = 0.125 \text{ in}^3$
- a high-shock environment, so $K_t = 3$
- a shaft with no keyway, so the shear stress limit is 6,000 psi.

Then, using the formula above, we find that the maximum torque we can safely carry is 6,000 x 3.14 x 0.125 / (16 x 3) = 50 inch-pounds.

Well and good, but how do we take this knowledge and do something practical with it like figure out the correct motor to use for expected speed and power requirements? The following section shows you how.

Determining Shaft Size for Given RPMS and Horsepower

Very often, the shaft-related question robot builders are most concerned about is how to determine the size of the shaft to be used, knowing the maximum rotational speed of the shaft and the maximum horsepower that will be developed. The correct steel shaft is determined by simply plugging numbers into the following equations:

$$\text{Steel Shaft Diameter (inches)} = 3.7 \times \sqrt[3]{\frac{\text{Horsepower}}{\text{RPM}}}$$

where there are no shock loads;

$$\text{Steel Shaft Diameter (inches)} = 4.5 \times \sqrt[3]{\frac{\text{Horsepower}}{\text{RPM}}}$$

where there are shock loads.

For example, let's say your robot's maximum speed is 10 MPH, the maximum horsepower transmitted is 2 hp per shaft, and the wheel diameter is 8 inches. First determine the RPM at maximum speed.

An 8-inch wheel travels 8 inches x π = 25 inches/rev, or 2.1 feet/rev. So, a shaft with an 8-inch wheel must turn at 10 miles/hour x 5,280 feet/mile x 1 rev/2.1 feet x 1 hour/60 min = 419 rev/min.

Now, we can just substitute into the formula for shaft diameter:

$$\text{Steel Shaft Diameter} = 4.5 \times \sqrt[3]{\frac{\text{Horsepower}}{\text{RPM}}}\text{ (shock loading assumed)}$$

$$= 4.5 \times \sqrt[3]{\frac{2\text{ hp}}{419\text{ RPM}}} = 0.75\text{ inches}$$

Bearings

Technically, a bearing is any machine component that holds or "bears" another component in place. Robot builders are most concerned about bearings that hold and support a rotating shaft. Bearings may be simply oiled holes in a chunk of metal through which the shaft is inserted, or they may be rings of precisely machined steel balls that allow the shaft to turn with little friction. There are many types of bearings. The most commonly used in fighting robots are journal bearings, pillow block bearings, and flange bearings. The correct choice for an application depends on many variables.

The first item to determine is the diameter of the shaft. Once this is determined, figure out whether light, medium, or heavy-duty designs are required, and work with a supplier to choose the right

9.1 Pillow block bearing

one. Finally, figure out if the bearing will rest on a horizontal surface (use pillow blocks) or a vertical surface (use flange bearings).

Bearings are made from various materials, including steel and cast iron. Experienced robot builders avoid cast-iron housed bearings, especially in supporting weaponry, as they often shatter upon impact.

9.2 Roller chain

9.3 Roller chain sprocket

Roller Chain and Sprockets

Roller chain and sprockets are one of the easiest and most versatile driveline systems for the beginning robot builder to tackle. In addition to allowing you to engineer your robot to travel at a desired speed and with adequate torque, they allow you to drive multiple wheel-connected shafts from a single driving shaft. A basic sprocket is a toothed circle of metal, specially designed to mate smoothly with a corresponding metal chain, called roller chain. Roller chain comes in various sizes. Common sizes for robot building are designated as numbers: 35, 40, 41, and 50. The larger the number, the heavier the materials from which the roller chain is fabricated.

Examining the smallest sprocket in the drivetrain, and noting the number of teeth on the sprocket and the speed at which that sprocket turns, determines the amount of power that can be transmitted by a roller chain. The following table provides horsepower ratings for Number 40 chain for various sprocket sizes and speeds.

Number of teeth in smallest drive sprocket	Small sprocket maximum RPM: 100	Small sprocket maximum RPM: 300	Small sprocket maximum RPM: 500
11	0.38 hp	0.96 hp	1.41 hp
15	0.53 hp	1.39 hp	2.09 hp
19	0.69 hp	1.80 hp	2.72 hp

As you can see, the fewer the number of sprocket teeth and the lower the chain speed that turns this small sprocket, the lower the horsepower transmission limit for the chain.

Why is this so? Because the horsepower transmitted by a chain is determined by the equation:

$$Power = Torque \times Rotational\ Speed$$

So, for a given horsepower rating, you'll find that the lower the speed, the higher the torque. The higher the torque, the more tensile stress placed upon the individual elements in the roller chain, ergo, the beefier the steel must be. For other sprocket sizes and chain speeds, consult an engineering handbook. The best handbook for basic mechanical design is the venerable *Machinery's Handbook* (Industrial Press) by Erik Oberg, Franklin D. Jones, Holbrook Horton, and Henry Ryffel.

Once the chain and sprocket sizes are selected, they must be mounted on the robot's frame. Positioning the chain in the robot is straightforward, but there are a few design considerations.

First, you must allow for chain tensioning. This allows for a greater tolerance in sizing the chain to fit between your sprockets and makes your robot more robust after it gets whacked during combat. To tension the chain, place one shaft on a bearing that fits into the robot frame in slotted holes. This allows you to slide the shaft so that the tension in the chain is right, even if things get moved a bit. Alternatively, you can insert a freewheeling sprocket called an "idler" to tension the chain.

Beginning builders often ask, "Why use a sprocket and chain, over, say, gears?" The answer is simple: Sprockets and chains easily take power from one shaft and place it where it is needed on another shaft some distance away. With gears, the driving gear and the driven gear must be in direct contact in order to transmit torque and power.

Diagram 9.4 shows how torque is taken off the motor output shaft and transferred to a shaft with a wheel attached. Gears require very precise alignment to one another and this may be difficult for the builder who is not a machinist or millwright.

9.4 Spur gears

The other important use for a sprocket and chain is to change speed and torque. The rotational speed from one toothed sprocket to another is determined by the ratio of the number of sprocket teeth on each. The following sprocket equation can be used to speed up or slow down the rotational speed of a shaft.

$$\frac{\text{Number of Teeth on Sprocket A}}{\text{Number of Teeth on Sprocket B}} = \frac{\text{Rotational Speed of Shaft B}}{\text{Rotational Speed of Shaft A}}$$

The amount of power that a chain drive transmits is provided by the torque-speed-horsepower equation:

$$\text{Horsepower Transmitted}_{\text{chain drive}} = \frac{\text{RPM x Torque (inch-pounds)}}{63{,}025}$$

Guidelines for Designing Roller Chain Drives

1. Choose the size roller chain you need based on the amount of torque you'll transmit. Number 40 chain can transmit

horsepower according to the chart above. Consult manufacturer's data for other chain sizes.

2. Determine the amount of speed reduction necessary between the motor shaft and the wheel shaft. With this in mind, choose the input and output sprockets according to the formula above. If your motor turns at very high speeds, and very large reductions in speed are necessary, do it in multiple stages.

3. Determine the approximate length of chain required by using the following formula:

Roller Chain Length = 2 x Shaft-to-Shaft Distance
+ Sprocket A Diameter x 0.5 x π + Sprocket B Diameter x 0.5 x π

4. Because a chain is made up of links with discrete lengths, you may not be able to size your chain to correspond exactly to the distances between shafts. However, half-links are available to slightly shorten the chain. Even so, you may need to include slots in the bearings holding the shafts in order to obtain a good fit.

5. Roller chain and sprocket alignment issues are not as critical as with gears, but great care must still be taken to align sprockets and chains.

Belts and Pulleys

Designing your robot drivetrain with belts and pulleys is very similar to designing one with chain and sprockets. The formulas for determining the speed ratios and the horsepower transmitted are the same.

Some robot designers prefer to work with chain drives and others with belt drives. The table below describes the differences between the two drive schemes. Different robot builders interpret these characteristics differently. For instance, slippage may not be desirable in terms of pushing another robot into a hazard, but it may protect the motor from locking up and burning out.

BELT DRIVE VERSUS CHAIN DRIVE

Attribute	Chain Drive	Belt Drive
Noise	Noisy	Quiet
Durability	Highly durable	Not as durable as chain
Alignment requirements	Precise	Less precise
Shock load tolerance	Medium	High
Slippage	No slippage occurs	Slippage may occur
Power transmission efficiency	Very high	Fairly High

Example 1

Suppose you already have a motor in mind, and you want to design a sprocket and roller chain drivetrain using this motor. The motor data sheet provides information on the motor's power, speed, and torque characteristics. The motor data sheet contains data obtained from a dynamometer test of the motor and tells us that at a design speed of 1,000 RPM the motor puts out 100 inch-pounds of torque. Let's choose drivetrain parts based on 1,000 RPM, 100 inch-pounds torque, and a desired forward speed of 9 miles per hour at these conditions.

First, choose a wheel size. You have access to some 6-inch diameter, good quality wheels and decide to use them. The table in Chapter 5 (page 53) tells us that a 6-inch wheel must turn at about 500 RPM to travel 9 miles per hour.

Since the motor shaft turns at 1,000 RPM, and we want the wheels to turn at 500 RPM, we need to use chains and sprockets or gears to obtain a 2:1 reduction ratio. At the same time, the torque to the wheels will be increased from 100 inch-pounds to 200 inch-pounds. Looking at the sprocket and chain vendor catalog, we can easily find sprocket diameters of 2.5 inches with 14 sprocket teeth and 5 inches with 30 teeth. This is slightly more than a 2:1 reduction, but close enough for your purposes. What chain is required? The table shown on page 102 shows that for Number 40 roller chain, the maximum horsepower transmitted for the smaller of the two sprockets (the 14-tooth sprocket) turning at 500 RPM is 2.09 hp. Horsepower is determined from this equation:

$$\text{Horsepower Transmitted}_{\text{chain drive}} = \frac{\text{RPM x Torque (inch-pounds)}}{63{,}025}$$

Substituting what we know into the chain drive equation, we get:

$$\text{Horsepower} = \frac{500 \text{ RPM x } 200 \text{ inch-pounds}}{63{,}025} = 1.58 \text{ hp}$$

Since 1.58 is less than the 2.09 horsepower upper limit for Number 40 roller chain, this is a workable design. Even if we increase the loading by 1.25 to account for shock loading (1.58 hp x 1.25 = 1.97 hp), it is still within design limits.

In actuality, the precise choice is predicated upon the power transmitted from the motor to the wheels, the number of starts and stops per day, the variability of the loading, and other factors. For your purposes, you simplified the calculation by ignoring the shock loads, and starts and stops simply added a safety factor at the end.

By knowing the motor characteristics and the desired speed and power at the wheels, you can choose a workable wheel diameter and then calculate the gear/sprocket reduction and choose the sprockets, shaft diameter, roller chain, and bearings.

The robot plans and blueprints show that the distance between the motor shaft and the wheel axle is exactly 10 inches. The next step is to determine the length of the drive chains.

From the blueprint, we know the shaft-to-shaft distance is 10 inches. We just sized the sprockets, and the diameter of motor shaft sprocket is 5 inches and the diameter of wheel axle sprocket is 2.5 inches.

The formula for sizing roller chain length for simple, two-element drives is:

$$\text{Roller Chain Length} = 2 \text{ x Shaft-to-Shaft Distance}$$
$$+ \text{ Sprocket Diameter A x } 0.5 \text{ x } \pi + \text{Sprocket Diameter B x } 0.5 \text{ x } \pi$$

Therefore,

$$\text{Chain Length} = 2 \text{ x } 10 \text{ in} + 5 \text{ in x } 0.5 \text{ x } 3.14 + 2.5 \text{ in x } 0.5 \text{ x } 3.14 = 31.8 \text{ in}$$

Example 2

Imagine you are designing the drivetrain for a low-riding robot. You plan to use a motor with a design torque at the motor shaft of 200 ounce-inches at 2,000 revolutions per minute. At this level, the motor data sheet shows the motor is generating about 0.5 horsepower. You want to use 8-inch wheels, and you feel a forward speed of 9 miles per hour is appropriate. How do you go about designing the drivetrain using sprockets and roller chain?

Step 1: Determine the required RPM of the wheel and wheel shaft. From the table of RPM and wheel diameters in Chapter 5, for 9 RPM, 500 RPM is about right.

Step 2: You need to reduce the motor speed from 2,000 to 500 RPM. Determine the number of teeth on each gear mounted on the wheel shafts. Substitute into the sprocket equation:

$$\frac{\text{Number of Teeth on Sprocket A}}{\text{Number of Teeth on Sprocket B}} = \frac{\text{Rotational Speed of Shaft B}}{\text{Rotational Speed of Shaft A}}$$

Teeth Sprocket A : Teeth Sprocket B = 4:1

The smallest sprocket you can find to fit your shaft has 9 teeth. The 4:1 reduction requires you to use a sprocket on the wheel shaft with 38 teeth. But according to the sprocket manufacturer's data sheet, a sprocket this size is about 7 inches in diameter. This is too large to fit within your robot's housing, so you use a double chain reduction instead. Diagram 9.5 shows how a double chain reduction works, and the speeds and torques at every important point in the drivetrain.

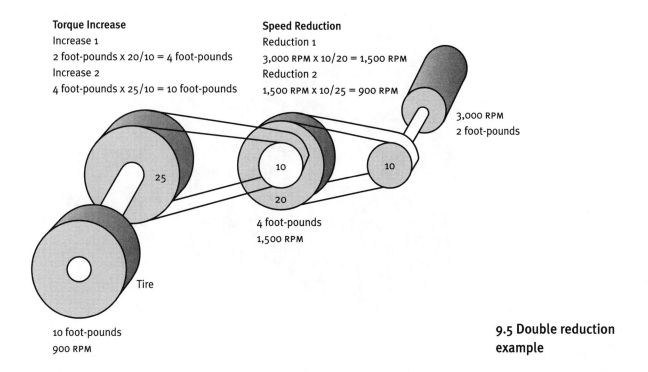

Torque Increase
Increase 1
2 foot-pounds x 20/10 = 4 foot-pounds
Increase 2
4 foot-pounds x 25/10 = 10 foot-pounds

Speed Reduction
Reduction 1
3,000 RPM x 10/20 = 1,500 RPM
Reduction 2
1,500 RPM x 10/25 = 900 RPM

3,000 RPM
2 foot-pounds

25

10

20

10

4 foot-pounds
1,500 RPM

Tire

10 foot-pounds
900 RPM

9.5 Double reduction example

Wheels

The wheels on your fighting robot have an important but tough job. They are the main point of contact between your bot and the fighting arena, and they are often the first place your opponent will try to attack. Strong, grippy, and tough wheels can give your robot a big edge.

Remember that bot building is all about trade-offs. Indeed, some of the desirable wheel qualities are mutually exclusive, so you must choose carefully for the best performance for your robot. Your wheels must be designed to get traction when accelerating, and be strong enough to withstand impact from the rams, spikes, and other nasty items your opponents use.

Wheels must also be fairly light, so the robot doesn't squander too much weight here, and must have hubs (where the axle shaft connects to the wheel) that are adaptable to the driveshaft. They shouldn't wear out too quickly and, finally, must be affordable for your building budget.

9.6 Wheels and tires

Traction

The traction you'll get from your wheels is based on two things: the weight of your robot and the coefficient of friction between the wheel and the floor. Weight is easy to understand, but what is the coefficient of friction? The coefficient of friction, or COF, is the quotient obtained by dividing the value of the force necessary to move one object over a surface at a constant speed by the weight of the object. The higher the COF, the more traction it will have. The COF between any two items can only be determined experimentally. Engineers use sophisticated testing equipment called tribometers to determine COF. Rubber and certain types of plastics have high COFs and are commonly used in robot wheels. Wheels made from less sticky materials are frequently perimetered with belts from vacuum cleaners or similar devices in order to improve traction.

Strength

Generally, robot wheels are designed for some other purpose, such as go-karts or handcarts. The wheels will be rated to carry a certain weight, so you'll need to compare the weight of your robot to the wheel's capacity rating. Using too light a wheel will often result in a fracture or deformation in shape, making your robot impossible to drive.

Connecting Wheels to Driveshafts

Handcart or wheelbarrow wheels are designed with a rotating bearing in the wheel hub. For driving wheels, the rotation must be eliminated because you want the wheel directly connected to the shaft. There are a number of ways that builders direct connect wheels to shafts: they use keys and keyways, they extend a strong pin through the wheel and shaft, and sometimes they directly weld the shaft to the wheel. Chapter 8 gives information on attaching circular weapons to shafts.

Many builders feel that rubber or polyurethane surfaces supported by aluminum or engineered plastic hubs provide a good combination of strength, traction, and machinability.

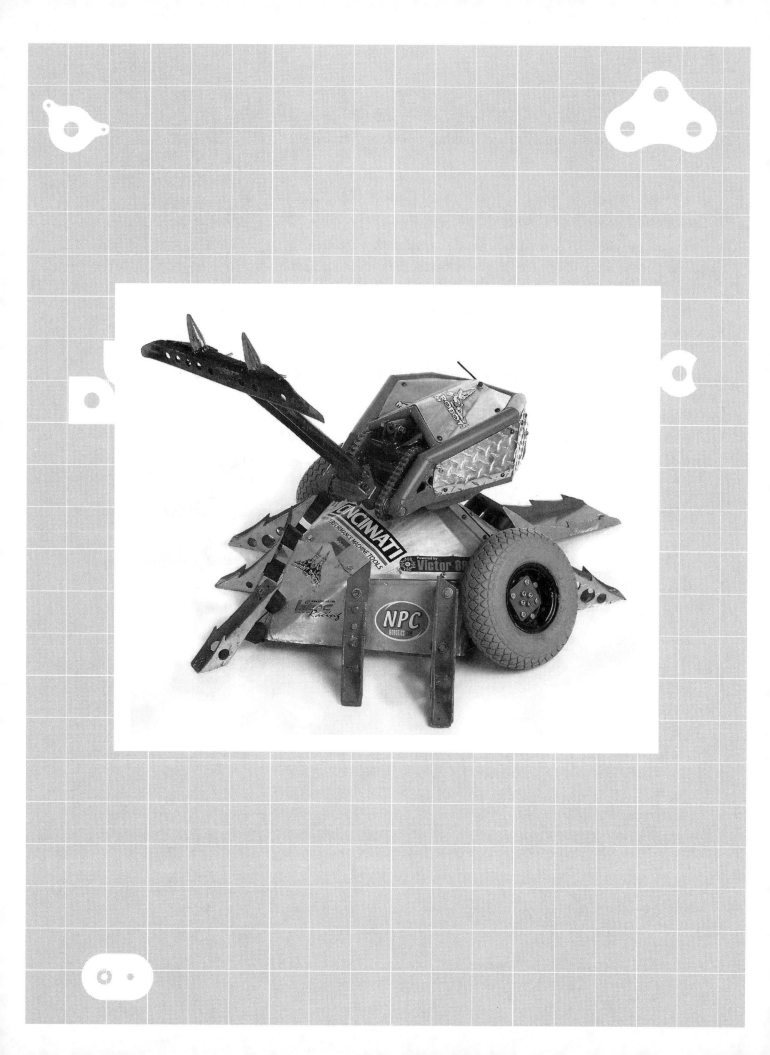

10.
MATERIAL PROPERTIES

Hang around a robot competition for a while and you'll probably hear the robot builders discuss the advantages of, say, 2024-T6 aluminum versus titanium alloys. They'll talk about the smell of a machine shop. They enjoy solving puzzles like determining how to choose the best frame materials in order to design a robot that is optimized in terms of resilience, weight, cost, and toughness.

As I've said before, robot building requires making choices between incompatible alternatives. This is especially true in selecting frame materials. A material that is very strong may be very difficult to fabricate. A very stiff and hard material may be impossible to weld. A material that seems strong in several material properties may be too expensive. Therefore, every bot builder needs at least some background in materials science and materials selection in order to choose the right stuff for the job.

Robot frame, weapon, and armor materials generally involve metals such as steel and aluminum, and exotic materials, like titanium. In addition, many builders use polycarbonate plastics and even composite materials such as carbon fiber because they are lightweight, transparent, and machinable.

Although fighting robot competitions have only been around since the early 1990s, robot frames have become highly evolved mechanical structures. Builders have a

Tazbot COURTESY BATTLEBOTS, INC., AND
DONALD HUTSON

variety of day jobs, like mechanical engineering, materials science, metalworking, and so forth. Some builders are experts in materials used for aircraft or bicycles, or in Hollywood-style, special-effects models, and they bring their expertise with them into this new sport. Their ideas are shared and copied, and many robots are now made from very sophisticated materials.

How to Choose Your Materials

Choosing the best material for a robot part depends on what you want that part to do. In general, the robot is composed of parts that hold other robot components (the frame), that act as protection from attacks by other robots (the armor), and that attempt to damage other robots (the weapon). Each part needs to do different things. Let's consider each part separately.

The Frame

The frame holds motors, wheels, and drivetrain components at precise locations in order to provide smooth-running and non-interfering drive motion. Therefore, frame parts should be very stiff and strong so the alignment of moving parts stays constant. Key frame material attributes are high stiffness, low density (lightness), and high tensile and compressive strength.

The Armor

Armor protects your robot during matches. It should be able to withstand great impact without failing. It is acceptable, or even desirable, for armor to deform, as long as it doesn't appear damaged to judges and spectators. The key armor attributes are toughness, low density, and resilience.

The Weapon

Tensile strength and hardness are key weapon material attributes—

the builder wants to be able to scratch and gouge opponents without breaking the weapon.

A beginning robot designer will often choose the strongest steel or aluminum alloy available within his or her budget and then start building. This is not the right approach. As noted above, strength may not be the design criterion for the part. Always match the material properties to the intended use.

Properties

To choose the best materials for your robot, start by understanding four important metallurgical properties: density, stiffness, ductility, and tensile strength.

Density

Density is the mass of a substance divided by its volume. For example, 6061 aluminum weighs 0.098 pounds per cubic inch. 4130 steel weighs 0.283 pounds per cubic inch, and 4-6 titanium weighs 0.160 pound per cubic inch. Two rules of thumb for comparing the density of frame materials are that titanium is about half the density of steel, and aluminum is about one-third the density of steel. A strong, light material is generally considered good; a heavy, weak material is bad. That's why you see far more robots constructed from aluminum and alloy steel than lead, sandstone, or fruitcake.

Stiffness

Material data sheets list the measured stiffness of a material by what is called the modulus of elasticity, or Young's modulus. This, like density, is reasonably easy to understand. Imagine a diving board, and think of stiffness as how much the board will deflect for a certain load. The higher the Young's modulus, the stiffer it is. While stiffness in armor is not necessarily a desirable property, stiffness is important in drivetrains where flexing may cause alignment problems.

Ductility

Ductility is a property that describes how much a metal will elongate or stretch before it breaks. Sometimes it is called elongation, which is the ability of a material to deform permanently without fracturing. Saltwater taffy has lots of ductility. A potato chip has none. When another robot hits yours, which is better? I'd go with the taffy.

Tensile Strength

Tensile strength is a very important property. It is measured by placing a material sample in a powerful materials testing machine that literally pulls the material apart and records the force required to do so, as well as the deformation to the material. Although chances are remote that somebody is going to attack your robot by trying to pull the beams and structural members apart, tensile strength is a very good indicator of how much impact your frame or weapon can take without breaking.

The Composite Properties: Toughness and Resilience

Generally, there are a couple of things you want out of the materials you use for your fighting robots: toughness and resilience. Commonly used words, certainly, but what do they mean in an engineering sense?

Resilience

Resilience is the ability to absorb energy from an impact and then release it. Rubber tires are resilient and window glass is not.

The best way to understand resilience from an engineer's perspective is with a diagram called a Material Stress-Strain Diagram. Stress on a part is the amount of force applied to the robot part divided by the cross-sectional area of the part normal to the direction of the force. Strain is a quantity that indicates how much a part lengthens under stress. It is measured by dividing the distance the material lengthens by its original length. Chewing gum exhibits huge strains before breaking, a Saltine cracker just about none.

Without going into too much detail, Diagrams 10.1, 10.2, and 10.3 show the amount of deformation or "give" a material will experience for a given stress level. Most robot-making materials have a stress-strain curve resembling the stress-strain diagram below. Here, the deformation will be proportional to the amount of stress placed on the robot part until the stress level reaches Point A. If the stress is removed, the member goes back to its original shape. All is as it was.

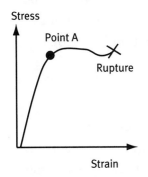

In a stress-strain diagram like this, tension is applied, and the elongation in the specimen is measured. A graph of stress (tensile force divided by the cross-sectional area) versus strain (elongation per unit length) is plotted.

10.1 Typical stress-strain diagram

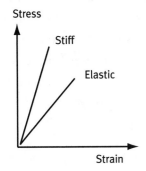

Stiff materials don't elongate much for a given stress.

Elastic materials do. When the stress is removed, the material snaps back.

10.2 Stress-strain diagrams for resilient and nonresilient materials

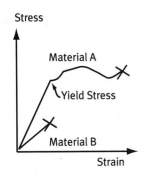

At some level of stress, steel "gives up" and is permanently deformed. This point is called the yield stress.

Material A takes a lot of stress up to its yield point, then deforms before breaking. This is a tough material.

Material B fractures easily, without absorbing much energy. Not tough!

After Point A, though, the material is said to yield, and the material is permanently deformed. Applying higher levels of stress after this point will result in permanent deformations.

10.3 Stress-strain diagrams for tough and nontough materials

A material is said to be resilient when the stress-strain curve looks like Diagram 10.2. It is said to be brittle when it looks like Diagram 10.3. Most robot builders want their robot to be able to absorb and release great quantities of energy from impacts, so resilience is a desired trait.

Material data sheets always provide tensile yield strength data. Stiffness is described in material data sheets as "modulus of elasticity" or "Young's modulus." In order to estimate resilience, divide the material's tensile yield strength by its stiffness. The higher this number, the more resilient it is. A material with high resilience might be a good material for armor and weapons.

Toughness

Toughness is the ability of material to absorb energy without breaking or fracturing. Really tough materials have lots of area under the stress-strain curve before breaking. This means you can beat the living daylights out of them and they will resist, resist, resist and then bend, bend, bend without breaking. Finally, after lots of high-level abuse, the part breaks. Toughness is a good characteristic for robotic armor and structural members. Some clever builders make armor out of tough stuff like steel, and then rubber-mount it, making it both tough and resilient, the best of both worlds.

PROPERTIES OF ROBOT FRAME, ARMOR, AND WEAPON MATERIALS

Properties	Density (1,000 kg/m³)	Young's Modulus (psi x 10⁶) / Modulus of Elasticity	Tensile Yield Strength (KSI)	Tensile Ultimate Strength (KSI)	Ductility (% Elongation)
1020 Mild Steel	7.85	30	24	44	22
4340 Alloy Steel	7.85	30	150	190	22
302 Stainless Steel	7.75	30	30	75	40
2024 Aluminum	2.6	10	42	65	20
6AL4V Titanium Alloy	4.5	15	130	135	54

Put another way, materials that can handle a lot of stress and have a good degree of ductility are said to be tough. You can estimate a material's toughness by multiplying its ultimate tensile strength by its ductility (elongation).

Based on the previous table, which materials would make the best robot frame or armor? (Remember, we're not factoring cost, weight, weldability, or machinability right now. We're just looking at material properties.)

Resilience Index = Tensile strength at yield / Young's modulus
1020 Mild Steel = 24 / 30 = 0.8
4340 Alloy Steel = 150 / 30 = 5.0
302 Stainless Steel = 30 / 30 = 1.0
2024 Aluminum = 42 / 10 = 4.2
6AL4V Titanium Alloy = 130 / 15 = 8.6

Toughness Estimate = Ultimate Yield Strength x Ductility
1020 Mild Steel = 44 x 22 = 968
4340 Alloy Steel = 190 x 22 = 4,180
302 Stainless Steel = 75 x 40 = 3,000
2024 Aluminum = 65 x 20 = 1,300
6AL4V Titanium Alloy = 135 x 54 = 7,920

Based on the figures above, titanium alloys have incredible toughness and resilience. Considering mechanical properties only, this small material sample ranks as follows: 6AL4V Titanium Alloy (best), 4340 Alloy Steel, 2024 Aluminum, 302 Stainless Steel, and 1020 Mild Steel (worst). Remember, robot building is all about trade-offs. The same properties that make titanium a great body material also make it very difficult to machine and weld. In fact, rating the machinability and weldability of these materials is almost the reverse of their material properties: 1020 Mild Steel (best), 302 Stainless Steel, 4340 Alloy Steel, 2024 Aluminum, and 6AL4V Titanium Alloy. In terms of cost, mild steel is cheap, followed by stainless, aluminum, alloy steel, and titanium alloy.

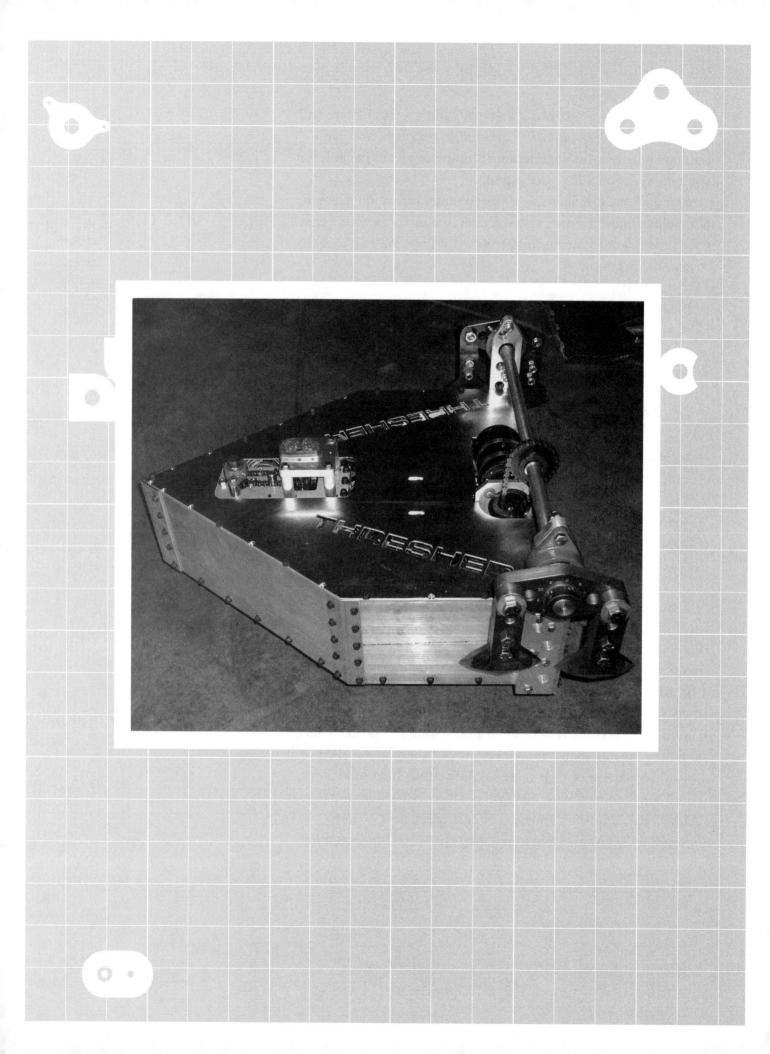

11.
BUILDING MATERIALS

Steel

There's something special but indescribable about working with steel. Like deer hunting and stock car racing, either you get it or you don't. Hobbyists have been making the metal chips fly with the help of a mill or a lathe for a long time.

Steel is easy to find at scrap metal companies, steel yards, and over the Internet. (Check out Appendix I for Web sites of metal suppliers.) It is available in many cross-sectional shapes: C-channels, L-shaped angle-irons, I-beams, pipes, drawn shapes (round, square, and rectangular), bar stock, and sheet. Sheet metal is available in a great number of standard thicknesses called gauges. High gauge numbers, like 28, correspond to thin sheets. Lower numbers, like 10 gauge, correspond to thicker sheets. Any sheet steel greater than $3/16$ inch is called plate. Bar stock includes round bars (threaded and unthreaded), squares, and flats. They generally come as thin as $1/8$-inch and go up from there.

Just like making chili, for example, there are different recipes for making steel, each with its particular characteristics. Most common is mild steel. It is widely available, fairly strong, inexpensive, and relatively easy to machine and weld.

The different ingredients in steel are called alloys and are used in various quantities and combinations to obtain

Thresher COURTESY TEAM NO QUARTER

Gauge Number	U.S. Standard Gauge (thickness in inches)	Mild Steel Weight (lbs/ft²)	Stainless Steel Weight (lbs/ft²)
8	0.1719	6.87	.....
9	0.1562	6.25	.....
10	0.1406	5.62	5.7937
11	0.1250	5.00	5.1500
12	0.1094	4.37	4.5063
13	0.0937	3.75	3.8625
14	0.0781	3.12	3.2187
15	0.0703	2.81	2.8968
16	0.0625	2.50	2.5750
17	0.0562	2.25	2.3175
18	0.0500	2.00	2.0600
19	0.0437	1.75	1.8025
20	0.0375	1.50	1.5450
21	0.0344	1.37	1.4160
22	0.0312	1.25	1.2875
23	0.0281	1.12	1.1587
24	0.0250	1.00	1.0300
25	0.0219	0.875	0.9013
26	0.0187	0.750	0.7725
27	0.0172	0.687	0.7081
28	0.0156	0.625	0.6438

special steel properties and performance. Alloying elements such as chromium, vanadium, and, most important, carbon, are mixed together with iron at the steelmaking plant to produce a huge number of steel alloys, all of which are made specifically to do different things.

For example, tool steel is a carbon alloy steel made for machining other types of metals, including mild steel. Obviously it is very hard, somewhat expensive, and difficult to machine into shapes. Some builders armor the undersides of their bots with tool or spring steel to provide protection from arena hazards, like kill-saws and the carbide-tipped blades of opposing robotic weapons.

Plate Thickness (in inches)	Mild Steel Weight (lbs./ft²)	Stainless Steel Weight (lbs./ft²)
$1/16$	2.56	2.59
$1/8$	5.12	5.19
$3/16$	7.68	7.78
$1/4$	10.24	10.37
$5/16$	12.80	12.96
$3/8$	15.36	15.56

Heat Treating

The hardness, toughness, and impact-resistance properties of steel, aluminum, and other metals can be enhanced by heat treating. Heat treating of metals is accomplished by placing a finished part in a furnace where it is brought to high temperatures at a specific heating rate, and in a controlled environment of air, oxygen, or nitrogen. After the part is brought to the desired temperature, it is cooled in water, air, or oil. This is done at a precisely described rate to attain the material properties desired.

Case hardening, annealing, tempering, and normalizing are examples of thermal processes used to make metals harder, softer, ductile, or tougher. Robot builders will often heat treat parts such as punches, rams, saw blades, hooks, and armor after they are fabricated to improve their material properties. Many robot builders take key parts to commercial heat treaters to improve their robot's battle performance. If you have specific parts you want to consider heat treating, check with a local company for advice and pricing.

Aluminum

Understanding aluminum alloys is not difficult, but choosing the right alloy for the job can be tricky. At the risk of being repetitive, choosing the right aluminum alloy is all about trade-offs. Some are quite strong, but won't weld. Some will weld OK, but won't machine worth a darn.

Most aluminum alloys are designated by a four-digit number. The first digit describes the principal alloying element added. In some cases, a heat-treating designation, such as T4, is appended to the alloy number. (The metallurgy involved in heat treating aluminum is beyond the scope of this book. Be aware, however, that you can sometimes improve the properties of machined aluminum parts by having them heat treated.) The most commonly used alloys of aluminum used in fighting robot fabrication are as follows:

1xxx series Aluminum of 99 percent or higher purity. 1xxx aluminum is used mainly in electrical and chemical applications. There aren't too many fighting robot applications for this aluminum alloy series.

2xxx series The 2 in this group means that copper is the main alloying substance. These alloys are often heat treated to obtain optimum properties in terms of tensile strength and durability. Sometimes called aircraft aluminum, 2xxx aluminum is used extensively for aviation applications. What's good for airplanes may also be good for robots, as heat-treated 2xxx aluminum can have material strength properties matching mild steel. Alloy 2024 is the best-known and most widely used aircraft alloy.

6xxx series Alloys in this group contain silicone and magnesium, making them heat-treatable. Some well-known alloys in this series include 6061 and 6063. Though not as strong as most of the 2xxx or 7xxx alloys, the magnesium-silicon alloys are easy to form into intricate shapes and resist weather and corrosion well. They have medium strength, so they are used for a lot of architectural applications. The aluminum bars and angles available at hardware stores and home stores are usually 6061 or 6063 aluminum products.

7xxx series Zinc is the major alloying element in this group and, when coupled with a smaller percentage of magnesium, results in alloys of very high strength. Usually other elements, such as copper and chromium, are also added in small quantities. The outstanding member of this group is 7075, which is among the highest strength alloys available. It is used in airframe structures and for highly stressed parts.

The strongest aluminum alloys are also the hardest to weld. The 2xxx and 7xxx alloys are very strong and very tough, which is why they are frequently used in aerospace applications. When you look at the wing of a modern airliner, you see thousands of rivets holding all of the aluminum wing parts together. Aircraft manufacturers use rivets because they just can't weld this type of aluminum. A robot built of, say, 2024 or 7075 aircraft alloy aluminum will have considerable strength, but it will be expensive and not weldable.

The extruded aluminum bars, rods, and angles commonly sold at hardware stores are almost always 6xxx alloy. These alloys provide medium strength and have medium weldability. A filler rod should be used when welding 6xxx aluminum.

Aluminum alloys are subdivided by a letter following the four-digit alloy number. The most common designation is T. An alloy number followed by T means the material has been heat treated. For example, 2024-T3 is aluminum that is aircraft-grade copper alloy and has been heat treated and cold worked. Heat treating aluminum improves the material performance and can be very worthwhile for critical parts. Again, if you decide that heat treating your aluminum parts is worthwhile, talk to some local heat-treating companies for advice.

Titanium

Titanium alloys are used in top-end fighting robot weapons and armor because they are light, tough, and have good fatigue and corrosion-resistance properties. But even more appealing is the strength, resilience, and toughness packaged in such a lightweight material. The specific weight of titanium is about half that of steel, and about 60 percent higher than that of aluminum. In tensile and sheet stiffness, titanium falls between steel and aluminum. But titanium's strength (80,000 psi for pure titanium and 150,000 psi and above for its alloys) is greater than that of many alloy steels, giving it the highest strength-to-weight ratio of any comparable metal.

There are two problems with titanium. First, it is difficult to machine and weld. You need to have some metalworking experience before tackling titanium projects. Titanium requires sharp tooling, slow cutting speeds, lots of cooling fluids, and precise material feed rates. While some builders try to machine it with hand tools, cutting titanium with a hacksaw is a crummy job and it is very hard on most drill bits. If you make holes in titanium with a drill press, keep the pressure constant, the feed rate slow, and use adequate cutting fluid.

The second problem is that titanium is very expensive. After acquisition and fabricating costs are factored, building with titanium can end up costing 20 times more than mild steel or 6xxx aluminum alloys.

Plastics

Many builders use tough, resilient plastics like Lexan (a brand name for polycarbonate plastic). They have decent strength-to-weight ratios and reasonably good material properties. One very good thing about them is that they are usually transparent, allowing spectators to see the interior workings of the robot.

Machining and Cutting Processes for Robot Builders

Different sorts of metals and plastics may be cut in various ways, depending on the material and its intended use. Heavy mild steel beams, for instance, are usually cut using a horizontal bandsaw, an abrasive cut-off saw, or a reciprocating power hacksaw. They may also be cut with an oxyacetylene cutting torch.

Plasma cutters, machines similar to arc-welders, liquefy a small, targeted pool of metal with a gas-shielded arc, then blow it away with an air jet. A high-performance water-jet cutter is similar in concept and is good for exotic materials such as titanium. If such equipment is beyond your budget and if a robot part is important enough, you can check the yellow pages for laser-cutting and water-jet firms. They will be happy to take on your work. And you can do a great deal of meaningful work without expensive tools simply by using a high-quality, handheld hacksaw and a sturdy vise.

Mild steel may be bent into various shapes without heating; it will take a considerable amount of deformation without failing. Metalsmithing forms and tools are also used for forming sheet steel into various shapes. While some of these are expensive and difficult to obtain, inexpensive tools sold for auto-body repair can be used to form robot parts.

If you're not interested in machining, you can still participate in robot building. Actually, there's no reason to feel that you must master all or even most metal fabricating skills to be a successful robot builder. Many builders simply use their creative and engineering skills to develop a workable design, and then send the plans out to a professional machine shop for fabrication.

Machine Shop 101

At this point, you're familiar with all of the components found in a warrior robot—motors, electronics, radio systems, materials, and so on. Now you can learn the fun part, which is putting it all together.

Robots come in a wide variety of different body shapes and sizes. Consequently, there is no single way to assemble everything. For example, some builders weld their robots together, some use bolts, and others may use something else entirely. Some builders cut structural members to size and build a frame. Others incorporate a strong sheet metal body, like an automobile's unibody construction, instead.

Machine Shop 101 is intended to provide some basic machine shop knowledge to help you start fabricating and assembling your bot. By applying good metalworking practice to your collection of parts, your robot will come to life, hopefully as a sleek, well-crafted statement of self-expression. There is no cookbook approach to fabricating a robot. There is no linear, universally accepted procedure to follow to ensure success. The best you can do is gain a basic appreciation of how mechanical things like this are built, and then put forth your best effort to bring your robot to completion according to your original design and intentions. The better your robot is machined, the better it will look, and the better it will perform.

With a few notable exceptions, no robot designer is strong in every aspect of engineering, and few robot builders are highly skilled in every machine shop process. Machine shop processes include drilling holes, cutting threads, sawing materials to length, milling shapes into metal, and so forth. Metalworking is considerably different than woodworking. For one thing, since metal is a lot harder than wood, metalworking takes longer and often requires more powerful tools. Therefore, it may be prudent to contract out all or part of the actual fabrication to professionals. There are many reputable machine shops that will work on jobs of this sort, at fair prices.

In order to perform the machining work yourself, or to communicate efficiently with your fabrication contractor, it is important to understand at least a few of the basic concepts and jargon of metalworking.

The most basic aspects of metalworking are covered here. If you decide to metalwork parts yourself, you must consult the operator's manual that comes with every metalworking machine. It provides instructions for its safe operation. There is a great deal to learn, so consult your manuals and books such as *Machinery's Handbook* (Industrial Press) for in-depth information. Be aware, machining metals and other materials involves sharp tools, lots of power, and the possibility of flying parts and metal chips. Read and observe the safety precautions associated with each shop machine.

Basic Machine Tools

The Drill Press

A drill press is preferable to a hand drill when the location and orientation of a hole must be controlled accurately. A drill press is composed of a sturdy base that supports a column. In turn, the column supports a table. The work—that is, the part on which you're working—can be supported on the table with a vise or hold-down clamps, or the table can be swiveled out of the way to allow tall work to be supported directly on the base. The height of the table can be adjusted by hand or with a crank, depending on the size of the machine, then held in place with a table lock. Above the table is a compartment that contains a motor. The motor turns the spindle at a speed controlled by either a variable speed control or a system of belts and pulleys. The spindle connects to a drill chuck to hold the cutting tools (drill bits, center drills, deburring tools, etc.). The cutting tools are moved up and down with a lever on the side of the drill press. The mechanical movement is counterweighted in order to obtain a smooth and precise motion when the spindle is lowered. Further, the drill press itself gives you a big mechanical advantage, letting you apply far more force on the bit than is possible with portable power drills.

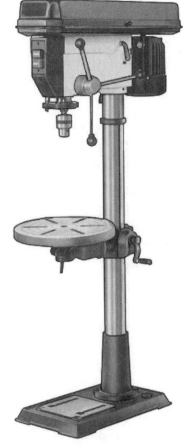

11.1 Drill press

The Lathe

The purpose of a lathe is to rotate a part against a hard metal-cutting tool. The part rotates and is pressed against the tool, which is positioned precisely by the lathe operator. The lathe is especially useful for fabricating parts and shapes that have a circular cross-section. The spindle is the part of the lathe that rotates. Various workholding attachments such as chucks and collets can be held in the spindle. As the spindle rotates, so does the workpiece. The spindle is driven by an electric motor through a system of belt drives and/or geartrains. The spindle speed is controllable by various adjustments on the lathe.

11.2 Lathe

The tailstock is a work holder used to support the end of the work piece. Alternatively, it can hold tools for drilling, reaming, or threading. It is supported in the desired position by steel rails or supports, called ways. The ways allow the lathe to accommodate different work pieces of different lengths.

The Mill

Milling machines are very versatile. They are usually used to machine flat surfaces, but can also produce curved or more intricate surfaces. They can also be used to drill, bore, cut gears, and produce slots. The type of milling machine most commonly found in a basic machine shop is a vertical spindle machine with a swiveling head. A milling machine removes metal by rotating a multi-toothed cutter that is fed or pushed into the work piece that is mounted to a moving table. The spindle can be fed up and down with a quill feed lever on the head.

Most commercial milling machines are equipped with motorized feed for one or more axes. Power feed is smoother than manual feed and therefore can produce a better surface finish. Power feed also reduces operator fatigue on long cuts.

Both lathes and mills are available in smaller "home-hobby" sizes that are fairly reasonable in price. While usually precise and accurate, they are limited in the size of the work they can handle.

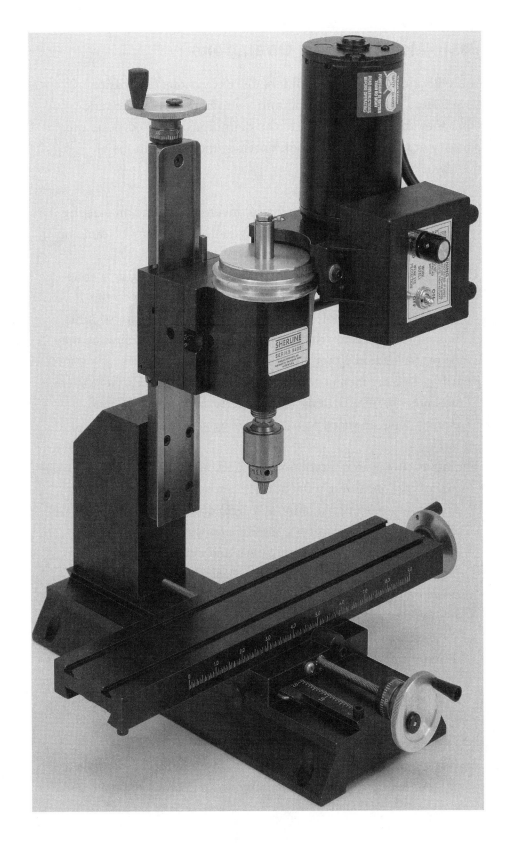

11.3 Mill. COURTESY OF SHERLINE, INC.

Basic Metalworking Operations

You will often hear robot builders discuss various fabrication terms. Here are some short definitions to help you understand what they mean. In addition to the definitions supplied here, the glossary at the end of this book provides more definitions of metalworking terms.

Boring Boring, in metal shop terms, means to enlarge an existing hole by using a single-point cutting tool. Boring may be done on either a lathe, mill, or drill press.

Countersinking Countersinking is the process of making a cone-shaped enlargement at the top of a hole. You might do this if you wanted to shape a hole to accommodate a screw head, so it fits flat and level (flush) with the part's surface. Countersinking may be performed on either a lathe or a drill press.

Drilling Drilling means cutting a hole in a work piece where there was none before. Drilling may be performed by using hand drilling tools or, more commonly, using a drill press, a mill, or a lathe.

Facing Facing is an operation performed on a lathe. It is the process of machining a flat surface across the face of a part by slowly removing metal chips at a right angle to the lathe axis. The lathe operator carefully moves a hard metal-cutting tool across the face of the spinning part, and cuts metal away perpendicular to the spinning axis. Basically, facing is used to decrease the length of a part.

Reaming Reaming is the enlarging of a hole to a precise and accurate diameter. A reamed hole is cut to a more precise interior diameter than a plain drilled hole. In order to maintain accuracy, reaming is always performed after a drilling or boring operation. Reaming is performed at much higher rotational speeds than most other shop operations. It is used to produce press fits, slip fits, et cetera.

Tapping The process of cutting angled, internal threads using a lathe or drill press is called tapping. Instead of a twist drill, a special tapping tool is used. It can also be done using a hand tool called, logically enough, a tap set.

Threading Threading is the process of cutting a screw thread on a part using a die handle, drill press, or lathe.

Turning When you use a lathe to remove metal from the outside of a work piece to form a cylinder, the process is called turning. It is similar to facing, except the cutting axis is parallel to the part instead of perpendicular to it. Turning decreases the diameter of a part.

Milling In conventional milling operations, a revolving cutting tool removes metal from a work piece. The spinning tool stays stationary, but the work piece moves through the movement of the table feed. The table position can be precisely controlled, providing very accurate part dimensioning.

Part Layout

In order to cut holes, lines, and other dimensions, the part must be marked so the machine tool operator can determine where the tool's cutting edges are to be applied. Marking lines, holes, and arcs on the work piece to guide the operator is called laying out the part. One good way to lay a part out on a piece of stock is to plot out a 1:1 scale drawing of the part and use spray mounting or glue to attach it to the part surface. If you do it this way, you can reduce errors made in transferring the part from the computer to the surface of the part.

If a full-scale drawing is not available, the part can be laid out by hand. This means lines and holes are drawn right on the part showing the areas where metal is to be removed. To facilitate accuracy, the machinist will often apply a thin coat of blue dye to the surface of the part. Then, lines are scratched in the dye using straightedges and squares. This makes it easy to see exactly where the metal is to be removed. The scribed line will be very thin, so you can make a reasonably accurate part by milling or drilling up to the line by eye.

Speeds, Feeds, and Other Important Data

There's a lot to know about cutting holes in metals. One of the most important things to know in order to cut metal safely and efficiently is the correct speed for drilling, milling, and other operations. Typically, a machinist consults a handbook that provides information on the correct feed and speed for different types of materials, hardnesses, thicknesses, and so on. One well-known handbook is *Machinery's Handbook* (Industrial Press), available in the reference section of most libraries and often in new and used bookstores.

In addition to feeds and speeds, a good handbook provides information on tap and die sizes, cutting fluids, gauge sizes, screw and bolt dimensions, material properties such as weight and hardness, and much more.

Welding and Fastening

A good thing about steel, and to a lesser extent aluminum, is that it is actually pretty easy to join pieces together. Here are some different ways you can do this.

11.4 Fasteners

The first and most obvious way to connect steel is with fasteners. Many people use bolts, which are quite strong and permit easy disassembly, or, less commonly, screws. Screws require a tapped or threaded hole in the metal. With screws, you only need access to one side of your work although you do need to invest in a tap and die set. Rivets, which are usually set by backing up the heads on one side with a tool while mushrooming the other side with a hammer, make a secure and permanent attachment. As far as glue and duct tape go, they're not strong and not pretty. I advise against using them since most people will see your robot as a "kludge," which means a poorly fabricated work.

In many cases, welding is the best choice for fabricating the frame of your robot. Several distinct processes have been developed, and each has its own advantages. Welding methods include: forge welding, oxyacetylene welding, shielded metal arc, and wirefeed metal inert gas welding.

Before we go any further, take heed: While this is probably extremely obvious, be aware that welding is dangerous if not approached properly. Welding involves very high temperatures, very bright light, and the possibility of hot metal sparks flying around you, as well as less predictable dangers. Use your head while welding. Don't light your cigarette from the flame of a gas welder, no matter how hip this looks on late-night television. Also, note that this introduction is not intended to provide the in-depth background needed to weld safely. Always refer to the directions and safety instructions that come with the welding equipment.

Forge welding is the original welding technique, developed by village blacksmiths over hundreds if not thousands of years. Forge welding is accomplished by bringing two pieces of steel to white heat (one can see the sparks emitting as the iron begins to burn). Then, while the steel is still white hot, the smith pounds the pieces together with a big hammer. As much as robot builders love big hammers and the idea of pounding metal, forge welding is generally for blacksmiths, not robot builders. There may be some forging application for making weapons, however.

A most versatile welding tool is the oxyacetylene torch. It takes more skill to master oxyacetylene welding than wire feed welding,

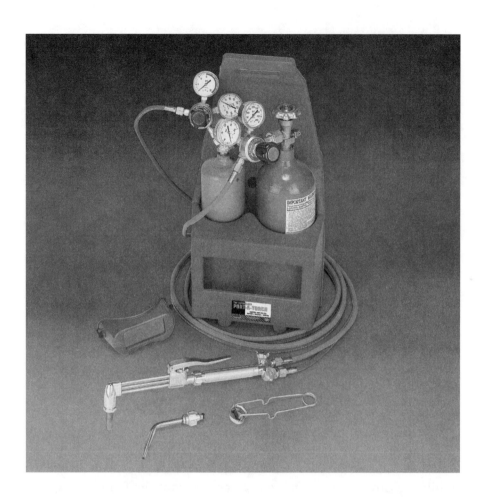

11.5 Oxyacetylene welding.

but it has several advantages. It is versatile—it works for welding, brazing, and soldering—and it usually comes with a cutting attachment that permits steel (but not aluminum or titanium) to be cut fairly easily, albeit roughly. Certainly, the hot flame of an oxyacetylene torch is dangerous and must be treated respectfully. Apart from the burn hazard, there is the also the danger of setting your workshop on fire or, possibly (although this is rare), causing an explosion by letting the volatile gas escape. Always follow the rules associated with the use and transport of acetylene. Also, be aware that the bright light of a gas welder or cutter flame can damage your eyes. Talk to experienced people and read up before you start welding.

The oxyacetylene welding process involves adjusting the gas torch so a small but distinct inner flame, called an oxidizing flame appears. With one hand, the welder takes a steel welding rod

11.6 **Shielded metal arc welding.** COURTESY OF LINCOLN ELECTRIC CORP.

appropriate for the task, carefully heats it past melting, and then guides the molten metal into the seam. Doing so joins the pieces together in the flowing puddle of metal. The torch must be moved carefully back and forth as it advances down the joint in order to keep the two sides molten at just the right rate, but not so hot they burn, and the filler metal must be added at precisely the right rate. Although the process is simple to describe, it takes lots of practice to get it right. Sheet metal is more difficult than angle iron, for example, because its thinness makes it easy to burn right through.

The other methods are variations of a general process called arc welding. In shielded metal arc welding, sometimes called SMAW but usually just called stick welding, a flux-coated rod is used as a combination of filler material and electrode. A heavy grounding clamp is placed on the object to be welded. Then the stick electrode is brought into close proximity of the work, establishing

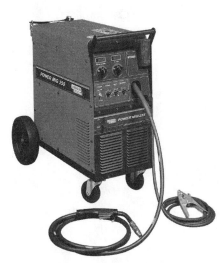

11.7 **Wirefeed welding equipment.** COURTESY OF LINCOLN ELECTRIC CORP.

the arc. By scratching the rod against the work—which incidentally you can't see, because the face shield is too dark to see through until the arc is established—the electrical circuit is established and welding begins. You're actually melting the steel rod with the electricity. Talk about a lot of current! Just like gas welding with a filler stick, the stick welder moves the electrode across the work and the puddle or bead of molten weldment is laid down joining the two work pieces together in a solid metal bond. The electrodes or rods come in many varieties. Most large home and hardware stores carry general-purpose welding rods that work well on mild steel, as well as rods for stainless steel, aluminum, and many other materials. Rods should be kept in a closed container since moisture in the air will cause them to deteriorate.

The easiest type of welding for the casual welder to master is wire feed, metal inert gas welding, or MIG welding. It is versatile, the equipment is not terribly expensive, and it is the easiest welding process for the beginner to use. Instead of using a filler rod electrode, a metal wire makes up a consumable electrode, and it is forced into the weld mechanically by simply squeezing a trigger on the welder's wire electrode holder. Jolted by very high amperages, the electrode liquefies and is sort of squirted into the weld seam. As neat as that sounds, it can be difficult to keep the wire flowing into the weld smoothly so the bead can get too thin or too thick. Small MIG units are available (see Appendix I) that run on 110-volt AC, single-phase, household electricity and are powerful enough, or so the manufacturers claim, to weld $1/4$-inch angle iron and most sheet metal.

MIG welding sounds great, but there are a few drawbacks. It is more expensive to set up stick or gas welders than if you're going to be welding a variety of materials, since an entire roll of wire must be purchased for each type of metal, instead of a stick or two. Welding thicker materials (non-sheet metal) is best done with a larger welder and these require at least 220 volts AC, which may be hard to find in your garage or workshop. Also, some users prefer oxyacetylene welding equipment because it is easily convertible to steel cutting and is usable in areas with no electricity.

Regardless of the welding process, you'll need to remove the excess material around the weld by grinding and sanding. Doing so creates the nice, smooth appearance of a seamless piece of metal. The tool most commonly used for this is the right-angle grinder, which comes in a range of sizes for running grinding wheels, sanding disks, and wire wheels.

12.
ROBOT PHYSICS

This chapter is for the Renaissance man (or woman) robot builder. You don't need to understand much physics in order to build a good robot. In fact, if you really disliked—even hated—your high school physics class, take heart; you can just skip this section. But if you'd like to understand the physics behind the motions and crashes of your bot, you might find this chapter interesting. Moreover, by applying basic science and engineering, you can optimize your designs and gain an advantage over your competitors.

Like a few other chapters in this book, think of the information here as a scientist's notebook. Instead of a linear exposition of knowledge, this chapter is a collection of facts and formulae. Some will likely be useful to your particular situation, and others may not.

Familiarity with basic physics facilitates the engineering of a powerful and effective robot weapon. This chapter is a review of some of the things you learned (or at least were supposed to learn) in your high school physics class. Physics is a broad science—it covers electricity, thermodynamics, atomic theory, and much more. The stuff discussed here comes under the general heading of Newtonian mechanics, which was developed by Isaac Newton in the 17th century. When engineers study mechanics, usually the first thing presented is something called kinematics, which describes and

Chaingor COURTESY JON VANDEVELDE

predicts the motion of things—particles, projectiles, and robots. Kinematics is a very large topic because it describes motion in both linear and rotational systems, and in one, two, and three dimensions. But take heart, because this chapter is only going to touch lightly on this avenue of physics, just enough to give you a start in figuring out how to develop an effective weapon.

Kinematics

Kinematics is the most elementary study in physics. It describes or predicts the motion of objects without being concerned with the forces that cause the motion. The basic measurements of kinematics are just two: time and distance. Physicists are a clever lot, and from playing with just these two quantities, they can write equations to tell you how to figure out all sorts of interesting things about how your robot will behave.

Velocity

Velocity is the rate of change of the position of robot over a period of time. In mathematical terms, velocity = distance/time. Velocity is a vector quantity, which means it has both a magnitude and direction. Therefore, for example, your robot moves with a velocity of 10 feet per second to the left, or 12 miles per hour north.

Acceleration

Acceleration is the rate of change in velocity (acceleration = velocity/time). Therefore, any time an object changes its rate of travel or its direction of travel, it is said to be accelerating. Acceleration is a vector requiring both magnitude and direction. For example, the acceleration vector that the earth's gravitation places on all objects at the earth's surface is 32.2 feet/second^2 downward, toward the center of the earth. So, for example, if a person falls off a ladder, his or her velocity will increase at a rate of

32.2 feet per second for each second while that person is in free fall. At the end of 1 second of free fall, the person will be traveling at a rate of 32.2 feet per second, or approximately 22 miles per hour. The person will have fallen approximately 16.1 feet in this 1-second interval. How do we know this? From the basic kinematics equations.

Basic Kinematics Equations

A physics textbook would normally derive a bunch of equations that describe relationships between distance, time, speed, and acceleration. The derivations involve using calculus and higher algebra. Lucky for you, this isn't a physics textbook, so it'll just list the equations and let you use them as needed. Normally, you identify the quantities you know, and solve for the ones you don't. For example, if you know the starting and ending velocity of your robot and the distance it traveled, there's a formula you can use to determine the time it took to cross that distance. These formulas are very handy!

BASIC KINEMATICS FORMULAS

A = acceleration
V_o = starting velocity
V_f = ending velocity
T = time
D = distance

Formula	To find:	Given these	Use this equation
1	Time Taken	A, V_o, V_f	$T = \dfrac{(V_f - V_o)}{A}$
2	Time Taken	V_o, V_f, D	$T = \dfrac{2D}{(V_o + V_f)}$
3	Acceleration	T, V_o, V_f	$A = \dfrac{(V_f - V_o)}{T}$
4	Acceleration	T, V_o, D	$A = \dfrac{2D - (2V_o \times T)}{T^2}$
5	Final Velocity	T, A, V_o	$V_f = V_o + A \times T$
6	Final Velocity	A, V_o, D	$V_f = \sqrt{V_o{}^2 + 2A \times D}$
7	Distance Traveled	T, A, V_o	$D = V_o \times T + \frac{1}{2} A \times T_2$
8	Distance Traveled	A, V_o, V_f	$D = \dfrac{(V_f{}^2 - V_o{}^2)}{2A}$
9	Distance Traveled	T, V_o, V_f	$D = \frac{1}{2} T \times (V_o + V_f)$

Example 1

If your robot went from a standing start to 12 ft/sec in a time of 6 seconds, what is its acceleration?

We know $V_o = 0$, $V_f = 12$, and $T = 6$.

So we'll use formula number 3,

$$A = \frac{V_f - V_o}{T}$$

$$A = \frac{12 \text{ ft/sec} - 0 \text{ ft/sec}}{6 \text{ sec}}$$

$$\text{Acceleration} = 2 \text{ ft/sec}^2$$

Example 2

A robot is coasting along the arena floor at a leisurely 3 feet per second. The driver provides full power to the big DC motors onboard and it accelerates at a rate of 10 ft/sec². To calculate the robot's final velocity after it travels across the 25-foot arena, plug what you know into the following formula and solve:

$$D = \frac{V_f{}^2 - V_o{}^2}{2A}$$

$$25 \text{ feet} = \frac{V_f{}^2 - (3 \text{ ft/sec} \times 3 \text{ ft/sec})}{2 \times 10 \text{ ft/sec}^2}$$

Rearranging,

$$V_f = \sqrt{509}$$

$$V_f = \text{approx } 22.5 \text{ ft/sec}$$

Example 3

A robot at one side of 30-foot arena accelerates at a rate of 8 ft/sec^2 from a standing start and crashes into the wall at the other end. To calculate the speed at wall impact, break this up into two parts. First, use equation 7 to figure out the time it takes to cross the arena.

$$\text{Distance} = (V_0 \times T) + \frac{1}{2} A \times T^2$$

Since $V_0 = 0$, then

$$30\,\text{ft} = 0 \times T + \frac{1}{2} \times \frac{8\,\text{ft}}{\text{sec}^2} \times T^2$$

Solving for time (T) tells you that the time it takes the robot to scoot across the arena is 2.7 seconds. Knowing this, then we can plug the time into equation 5 and obtain:

$$V_f = V_0 + A \times T$$

$$V_f = 0 + 8\,\text{ft/sec}^2 \times 2.7\,\text{sec} = 21.6\,\text{ft/sec}$$

Newton's Laws

Newton developed three laws of motion that relate kinematic phenomena to forces. Any push, ram, hammer blow, or robot impact involves forces. Forces are measured in physics according to their effects. According to Newton's Second Law,

$$\text{Force} = \text{Mass} \times \text{Acceleration, or just } F = ma.$$

The amount of mass something has, multiplied by the amount of acceleration it has or imparts, constitutes the force. If we apply Newton's laws and add a few basic ideas like friction, momentum, and torque, we can do some serious engineering on our robot. Newton's laws, paraphrased for robot builders, read like this:

- **First Law:** A robot does not change its state of motion except under the influence of an external force. That is, there is no change in the velocity of a body (neither in magnitude nor in direction) unless some force acts on that body.
- **Second Law:** The net force applied to and by your robot is equal to the mass of an object multiplied by its acceleration.
- **Third Law:** For every force applied by your robot, there is an equal and opposite force back on it. Therefore, if your bot bangs into a wall, the wall bangs back on your robot equally hard. The pusher and the pushed, the hammer and armor, both experience forces of the same magnitude but of opposite direction.

Let's add a few more basic ideas to round out the fighting robot physics section—ideas like torque, momentum, and energy, among others. Then we can look at some examples of what we can do in terms of engineering fighting robots.

Torque

A torque is a twisting force. It is a force that tends to induce rotary motion rather than straight-line motion. Torques are typically measured in foot-pounds. So if we pull on a one-foot-long wrench with a force of 100 pounds, we exert 100 foot-pounds of torque on the nut. A 50-pound pull using a 2-foot wrench generates the same torque.

Friction

Friction is a special kind of force produced by two bodies that are in contact. If a robot is at rest on the arena floor and another robot tries to push it, its efforts are resisted by what is known as static friction. Once it gets sliding it takes less force to keep it moving. If you stop pushing it, it will soon stop moving. The retarding or stopping force is known as dynamic friction.

Momentum

Momentum is the product of mass and velocity (momentum = mass x velocity). Momentum is a vector quantity just like velocity, acceleration, and force. The great thing about momentum is that it is always conserved on impact. That is, the total robot momentum before a collision is the same as the total momentum after the collision. Knowing this, you can compute the velocities of robots *after* a collision by knowing their speeds and directions of travel *before* the impact.

Energy and Power

Energy is the capacity to do work. In Newtonian physics, the energy of a body is computed in two ways, either by computing its kinetic energy:

$$KE = \frac{1}{2} M \times V^2$$
(M = mass, and V = velocity in linear systems)

or

$$KE = \frac{1}{2} \text{ Moment of Inertia (MOI)} \times V^2_a$$
(V_a = angular velocity in rotating systems)

or by computing its potential energy due to its height above the ground:

$$PE = M \times g \times h, \text{ where } g = \text{earth's acceleration}$$

Note that mass is not the same as weight. To convert weight to mass, divide an object's weight by the earth's gravitational constant, 32.2 ft/sec^2.

$$\text{Mass (pounds mass)} = \text{Weight (pounds weight)} / 32.2$$

Computations involving kinetic energy are not very tricky, but they are incredibly informative. For example, if a robot crashes into a solid arena wall at 20 miles per hour, the kinetic energy dissipated in the crash = ½ M x 20² = ½ M x 400 = M x 200. But if it crashes into the wall at 40 miles per hour, the energy dissipated in the collision is ½ M x 40 x 40 = ½ M x 1600 = M x 800. So, four times as much energy is involved in the second crash as in the first (M x 800/M x 200 = 4). The speed has doubled but the energy has quadrupled!

Kinetic energy calculations are also used to determine how effective a robot weapon will be. The greater the energy the weapon produces and transfers to its opponent, the more effective it will be. The following sections discuss this idea in some detail.

Power is the rate at which energy is used. For linear (non-rotating) systems it can be calculated from force and velocity by using the following equation:

$$Power = Force \times Velocity$$

For rotating systems, like an electric or gas motor, the formula is similar:

$$Power = Torque \times Angular\ Velocity$$

G Forces

The term "g force" is convenient for describing accelerations due to impacts. One "g" is an acceleration equal to that generated by the earth's gravitational field—32.2 feet per second per second. A 100-pound robot component acted on by a 2-g acceleration experiences a force equal to 200 pounds. This force acts in the direction of the acceleration.

During high-speed collisions, accelerations in the range of 5 to 25 g's are not unusual. For example, if a 150-pound robot is acted on by a 10-g stopping force during a collision, then a force of 1,500 pounds acts on that robot. This is high enough to break critical parts. Note that while a g value is really acceleration, it is sometimes looked at as though it were a force multiplier. This is not completely kosher, technically speaking, but it's not a problem if it is clear on which item the force acts. The good thing about using g's is that they allow you to simply multiply weights by g loads to determine forces caused by acceleration or deceleration.

Pressure

Pressure is force per unit area. Thus, if a 10-pound weight has a contact area with another body of 1 square inch, the pressure is 10 pounds per square inch.

Stress

Stresses, like pressure, are also forces per unit area. However, while pressure acts on robots from the outside, stresses are found in the parts of the robot. If you pull on both ends of a steel shaft of 1 square inch cross-sectional area with a force of 100 pounds, you will generate a tensile stress inside the shaft of 100 pounds per square inch (psi). If you twist across the long axis of the shaft, you'll generate a shear stress.

Fighting Robot Physics Examples

Example 1

Suppose a 40-pound robot were to crash into a second robot weighing 45 pounds, and you were interested in determining the forces the robots produced due to the impact. How is this done? First, take a few simple measurements: The moving bot is gliding along, unpowered, at 10 miles per hour (14.5 ft/sec), the second bot is at rest. For reference, note that

$$1 \text{ mph} = \frac{5,280 \text{ ft}}{\text{hr}} \times \frac{1 \text{ hr}}{3,600 \text{ sec}} = 1.47 \text{ ft/sec}$$

The momentum of the system before the collision is

$$\text{Momentum} = M \text{ (moving robot)} \times V \text{ (moving robot)} +$$
$$M \text{ (standing robot)} \times V \text{ (standing robot)}$$
$$= 40 \times 10 + 0 \times 45$$
$$= 4{,}000 \text{ lbs-miles/hr}$$

According to Newton, in a collision of this type, momentum is conserved. So, the following relation holds true:

$$\text{Momentum before crash} = \text{Momentum after crash}$$
$$4{,}000 = 40 \times V + 45 \times V = 4{,}000/1{,}800 = 2.2 \text{ mph } (3.2\text{ft/sec})$$

where "V" is now the post-impact velocity of the robots. (For simplicity, assume here that both robots are moving with the same velocity after impact.)

The first robot loses 7.8 mph (11.5 feet/second) due to the crash.

Interesting enough by itself, but what else can be found? Say the first robot has a hardened steel ram, and the second robot has fairly soft aluminum armor. When the robots collide, the second bot's armor dents by about 3 inches during the impact, and it slides about 9 inches. The distance through which the retarding force acts is therefore 12 inches. Therefore, the second robot travels about 1 foot during the impact with an average velocity of about 11.5 feet/second. The duration of the impact can be estimated as from the basic equations of kinematics:

$$\text{Distance} = \text{Velocity} \times \text{Time}$$
$$\text{Time} = \text{Distance/Velocity}$$
$$T = \frac{1 \text{ ft}}{11.5 \text{ ft/sec}} = 0.087 \text{ sec}$$

So, the first robot decelerates from 14.7 feet/second to 3.2 feet/second in a time of 0.087 seconds. Its average deceleration is:

$$\frac{(14.7-3.2) \text{ ft/sec}}{0.087 \text{ sec}} = \frac{132 \text{ ft/sec}}{\text{sec}} = 4.1 \text{ g's (negative)}$$

Using the g force multiplier technique, the average force acting on the robot then is 45 lbs x –4.1 g's = –180 pounds of force.

Example 2

Using the above example, what happens if you substitute a sharpened spike with a tip area of $\frac{1}{25}$ square inches for the ram? The spike impacts the second robot with a pressure of 180 pounds force / 0.04 in² = 4,500 psi. The pressure is high enough to possibly cause damage or deformation to the aluminum armor.

Example 3

The builder of the robot Acellerondo tells you that the four DC motors on his heavyweight (322-pound) robot are sized such that it reaches a speed of 18 MPH after going just 5 feet. Can you validate his claim? How big would each of his motors have to be in order to do this?

First, determine the time it would take Acellerondo to go from a standing stop to 18 miles per hour (26.5 ft/sec) in 5 feet using kinematic equation number 2: $T = 2D/(V_0 + V_f)$.

$$\text{Time} = \frac{2 \times 5 \text{ feet}}{0 \text{ ft/sec} + 26.4 \text{ ft/sec}} = 0.4 \text{ seconds.}$$

So, Acellerondo would need to accelerate from 0 ft/sec to 26.4 ft/sec in a time of 0.4 sec. Then, its average acceleration from Equation 3, $A = (V_f - V_0)/T$, must be:

$$A = (26.4 \text{ ft/sec} - 0 \text{ ft/sec}) / 0.4 \text{ sec} = 66 \text{ ft/sec}^2$$

which is about a 2 g-force acceleration. Therefore, the average force acting on Acellerondo is 324 lbs x 2 g's = 624 pounds force. Horsepower is calculated by using the following equation for a nonrotational device like this robot:

$$\text{Power} = \text{Force} \times \text{Velocity}$$

You just determined that the force is acting on the robot is 624 pounds.

Assume constant acceleration. The average speed during acceleration is (Ending speed + Beginning speed) / 2.

$$\text{Average speed} = \frac{26.4 + 0}{2} = \frac{13.2 \text{ ft}}{\text{sec}}$$

The power required is determined by multiplying force by velocity:

$$624 \text{ lb}_f \times 13.2 \text{ ft /sec} = 8{,}237 \text{ ft-lb}_f/\text{sec} = 15 \text{ horsepower.}$$

So, from this analysis at least, it is possible, assuming Acellerando has four or more 5-horsepower motors working at better than 80 percent efficiency. It's doubtful, though, that you'd actually get anything near that much power delivered to the wheels, since there are losses throughout the system due to friction and heating, but it is in the ballpark.

Wrapping Up

The point of all this is that you can perform considerable engineering analysis on your proposed robot by simply applying basic physics. You can engineer your robot to the speed, weight, acceleration, and material strength characteristics you need to handle the force, torque, and momentum requirements you calculate.

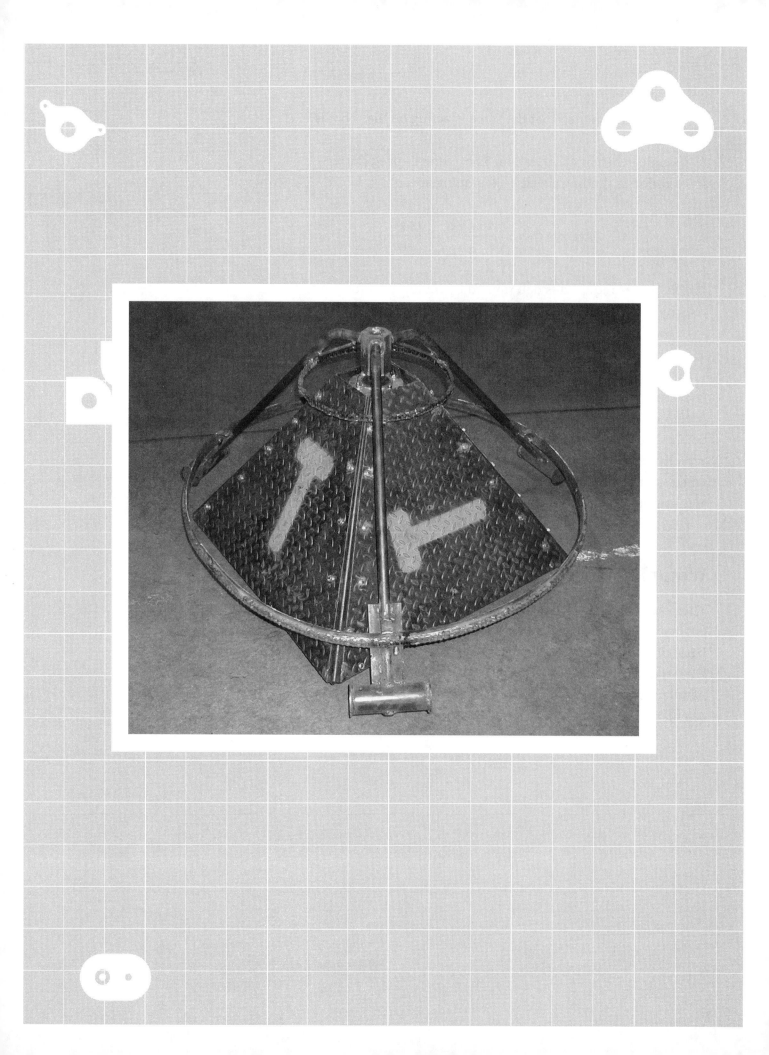

13.
ROTATING KINETIC ENERGY WEAPONS

For robot weapons, it all comes down to kinetic energy. Kinetic energy is a term physicists use to describe the amount of energy that's "in" something by virtue of its mass and motion. Getting beaned by a Nolan Ryan fastball is obviously more painful than being hit by Uncle Bob's whiffle ball. The more speed and mass a moving object has, the more kinetic energy it possesses.

Kinetic energy weapons are the overall categorization for such nasty items as swinging hammers, revolving drums, and, very frequently, spinning disks. (Large saw blade weapons, while technically kinetic energy weapons, are discussed in Chapter 14.) Spectators and judges respond favorably to kinetic energy weapons because they are visually very dramatic and, if done right, very effective. This section provides information about building effective kinetic energy weapons.

The main concept in kinetic energy weaponry is to build a durable striking object mounted on a heavy flywheel and quickly accelerate it to high speed using an electric or gas motor. Then, the robot driver guides the fast-moving, high-energy weapon and strikes a vulnerable part of the robot opponent. The basic idea is to make the weapon transfer its kinetic energy to the opponent's robot in a most destructive way!

Thor COURTESY TEAM NORSE

155

Here's the secret to why kinetic energy weapons are so effective: they allow you to accelerate your weapon for far longer than you can accelerate your robot in the arena. If you take a long across-the-arena run at another robot, you can accelerate only until you hit the other robot or the distant wall. But a spinner can just keep accelerating and accelerating, all the while building up kinetic energy in the spinning flywheel.

Physicists quantify the amount of kinetic energy in an object by using a quantity called foot-pounds or, in the metric system, joules. For simplicity, think of a spinning object with a great deal of energy being brought to a stop by striking another object. When this takes place, the spinner transfers its energy to the hit object, which absorbs the energy by being smashed, deformed, ripped, or flung around the arena. The kinetic energy is transferred from the spinner to the struck object.

Designing a Spinner Robot

A spinning weapon is typically a large steerable flywheel. (A flywheel is a rotating object that stores energy by virtue of its mass and the speed at which it spins. It is much like a mechanical storage battery.) An electric- or gasoline-powered motor turns a shaft that is mechanically connected to the robot's exterior, which is heavy and hard. The driver attempts to make direct contact with the robot opponent and transfer energy from the spinning exterior to the opponent. Theoretically, the large energy quantities transferred will damage connections to the opponent's drive system, rendering it inoperable, and the spinner chalks up a TKO.

Spinner Physics

In order to determine the power and efficacy of your spinning weapon, two equations prove very useful.

Equation 1: Total Kinetic Energy of the Weapon:

$$\text{Kinetic Energy} = \frac{1}{2} \times \text{Moment of Inertia} \times \text{Angular Velocity}^2$$

The moment of inertia (MOI) is a somewhat difficult quantity to describe, as it can only be rigorously defined by using calculus; feel free to dig up a beginning calculus book if you want. If you don't want to, just imagine you divided up a big object into little chunks, and then multiplied each chunk's mass by the square of each chunk's distance from the object's rotational center. When you sum up all of those products, you get an approximation of the moment of inertia. Suffice it to say, the moment of inertia of an object is a consequence of both the object's mass and geometry.

The rotational speed of an object is usually measured in revolutions per minute. However, most kinetic energy equations want you to use a dimensionless quantity called radians per second instead. To convert RPM to rad/sec, divide RPMs by approximately 10:

$$\frac{\text{Radians}}{\text{sec}} = \frac{\text{RPM}}{10}$$

Confused by moments of inertia and radians? Here's an example to make things more concrete. Imagine your robot's weapon is approximated by a thin-walled spinning cylinder as shown in the equation below. From an engineering handbook, we find that the moment of inertia of a thin-walled cylinder is calculated from this equation:

Equation 2: Moment of Inertia (spinning cylinder)

$$\text{MOI} = \text{Mass of Cylinder} \times \text{Radius}^2, \text{ or } \text{MOI} = M_{\text{cylinder}} \times \text{Radius}^2$$

Assume your cylinder has a 2-foot diameter. So it has a radius of 1 foot. Now measure both the rotational speed of the motor at full power by using a strobe or similar measuring device, and the weight of the cylinder using a scale. You find that:

Rotational Speed = 300 RPM = approx. 30 rad/sec
Weight = 100 pounds

So, substituting what you measured into Equations 1 and 2, you'll find that the total kinetic energy in the spinning flywheel weapon is:

$$KE = \frac{1}{2} \times \frac{100 \text{ lbs}}{32.2 \text{ ft/sec}^2} \times (1 \text{ ft})^2 \times \left(\frac{30 \text{ rad}}{\text{sec}}\right)^2 = 1{,}400 \text{ ft-lbs}$$

If you like to think in metric terms, you can convert foot-pounds to the metric equivalent to obtain a kinetic energy of 1,900 joules.

Is this a lot of energy? Imagine that your robot was able to get underneath its opponent and strike it such that all 1,400 foot-pounds of energy were available to heave it straight up into the air. (Unlikely, certainly, but we're just trying to make a point.) The hit robot would go up and up until all of the kinetic energy of the strike was converted into potential energy. The conservation of energy principle tells us that

Equation 3: Kinetic Energy (spinner) Before Collision = Potential Energy (hit robot) After Collision
The potential energy is determined from this equation:

Equation 4: Potential Energy = Weight x Height
Assume the hit robot also weighs about 100 pounds. Its mass is 100 pounds (weight) divided by g, or 32.2. Then substituting:

1,400 foot-pounds = 100 lbs x Height

Solving for height,

Height = 14 feet

That's enough to bounce the other robot off the ceiling!
If instead of a spinning cylinder, your robot consisted of a spinning hemisphere, then the only change would be in the

moment of inertia calculation. The MOI for a spinning hemisphere is found in all engineering handbooks and is:

$$\text{Moment of Inertia}_{\text{hemisphere}} = \frac{2}{3} \text{ Mass of Cylinder x Radius}^2$$

Other moment of inertia calculations for other geometries are easily found in engineering handbooks.

Look carefully at the equations for kinetic energy. You'll see the amount of kinetic energy is directly proportional to the spinner's weight and increases by the square of its rotational speed. You can boost energy levels quickly by simply spinning faster. So, to make a more powerful weapon, just obtain a bigger motor.

But there are a couple of problems in doing so. First, the weapon motor must be powerful enough to spin the structure at the speed desired, and bring it to speed quickly enough in a battle situation. Powerful motors cost more and weigh more. Second, the bearings, shafting, drivetrain, and spinner itself must be strong and thick enough to handle the stresses placed on them by the higher rotational speed and the resulting linear impacts.

Supposing that the robot described earlier was built substantially enough, and the weapon motor powerful enough, you could increase the speed by 50 percent, to about 45 radians/second. Then the kinetic energy equation changes to

$$\text{Kinetic Energy} = 0.5 \times 100 \text{ lbs}/32.2 \text{ ft/sec}^2 \times 1 \text{ ft}^2 \times (45 \text{ rad/sec})^2$$
$$= 3{,}150 \text{ foot-pounds} = 4{,}270 \text{ joules}$$

This will really wallop an opponent's armor. Just for fun, let's calculate how high 3,150 foot-pounds would raise a 100-pound robot.

$$\text{Energy (potential)} = \text{Weight x Height}$$
$$3{,}150 = 100 \times \text{Height}$$

Solving for height:

$$\text{Height} = 31.5 \text{ feet (!)}$$

Safety Note

Everything that makes a spinning weapon effective also makes it dangerous. At a rotational speed of 1,000 RPM—which is not unheard-of from robots from top builders—the circumference of a 2-foot diameter robot experiences over 300 g's. This can easily cause problems—like complete destruction of the bot, with shrapnel—if inadequate walls, shafts, or bearings are used. Use great care in designing, building, testing, and using spinning weapons. They can be nasty!

Determining Motor Requirements

Now let's determine the size of the motor required to develop this level of kinetic energy. Suppose you specify that the spinner needs to go from stopped to full speed in a maximum of 2.5 seconds. How do you calculate the size of the motor required?

First, convert joules or foot-pounds to something more familiar for motor calculations, namely, horsepower-hours.

Equation 5: 1,980,000 foot-pounds = 1 horsepower-hour

If you ran a 1-horsepower motor for 1 hour, it would consume 1 horsepower-hour, or 1.9 million foot-pounds of energy. You now have what you need to determine the size of the motor required to produce 3,150 foot-pounds of energy within the 2.5-second spin up time requirement.

From Equation 5, you can calculate that 3,150 foot-pound = 0.0016 horsepower-hours. There are 3,600 seconds in an hour. So the energy produced by a 1-horsepower motor in one second is simply 1/3,600 of a horsepower-hour. You need to obtain 0.0016 horsepower-hours in 2.5 seconds. So the conversion calculation to determine the energy requirement is:

0.0016 hp-hrs x 3,600 sec/hr = 5.76 hp-sec

You're getting close now. You need to produce 5.76 hpw-seconds in a 2.5-second interval, which is the time interval you decided would be acceptable for going from stopped to full speed.

Horsepower is the time rate of providing energy, that is, the amount of energy produced per unit of time. Divide the energy required by the time required and you obtain:

5.76 hp-sec/2.5 second spin up interval = 2.4 hp

This is a best-case number. There will be transmission losses, friction, and motor inefficiencies, so figure on at least doubling the motor requirement, and use a 5-horsepower motor to get what you need.

One Last Thought

Always remember Newton's Third Law of Motion: Every action has its own and opposite reaction. When you deal a blow to another bot, you're applying one to your own as well. You have the advantage in where and how it gets applied, but your robot must still be built to handle this type of force. Spinning robots require a lot of optimizing and toughening so when you crash into your opponent, you don't get the worst of the transaction.

In the next chapter, you'll use the same kinetic energy ideas to determine how much "punch" the other types of robot weapons have.

14.
MORE ROBOT WEAPONRY

The basic, overarching purpose when building a warrior robot is to build it with the ability to do substantial damage to another robot. As one can imagine, this is a broad subject, and it provides a rich, fertile ground for exercising the most mischievous facets of the builder's imagination. But unlike a bar fight, there are rules. In fact, a typical warrior robot tournament is more tightly regulated than a Taliban square dance, but most of the time your good ideas can be incorporated. This chapter looks at more fighting robot weaponry and touches on tried-and-true technologies builders have used to kick someone's bot.

Fighting robots can be categorized by the type of weapons they employ to wreak destruction and damage on their opponents. In fact, a whole taxonomy of robots can be built based on the weapons and methods used to score points and knock out opponents in the ring.

Some robots disable their opponents by cutting through armor to make minced metal out of the critical drive and electrical gear found inside. Others disgorge huge amounts of energy to rotating flywheels and then bang the energy-laden spinner into opponents in an effort to shock load the opponent into submission.

Still other robots make use of electrically or pneumatically induced linear motion to flip the competitor head over heels and knock it out in the process. The

Psychotic Reaction COURTESY
TEAM KONTROLLED KAOS

following is a description of some of the more common types of
robotic weaponry.

BMWs

BMW stands for battery, motor, and wheels. When you have a
BMW, that's what you have: the raw power of mighty engines with
grunting torque encased in a hardy armor shell. BMWs use the
momentum developed by their own mass and speed to simply
smash into their opponent. They may have rams or wedges in the
front, or possibly steel spikes, but these are the simplest and most
straightforward of all fighting robots.

A BMW has no spinning disks to balance and adjust, no tungsten-
carbide cutting wheels that might throw a drive chain. The builder
of a BMW simply puts all his or her building effort, money, and
weight into big motors, big wheels, and big batteries. The robot
develops as much torque and speed as it possibly can given its
weight class restrictions, and then muscles and pushes opposing
robots into the arena walls, sides, and, most of all, the arena
hazards. This can be very effective, but success depends on driving
skills and the ability to absorb punishment as well as dish it out.

Besides pushing opponents into arena hazards, another strategy for a BMW is to transfer its kinetic energy (that's the energy it possesses by virtue of its speed and mass) to its opponent in a way that is relatively painless for the BMW but really destructive to the other robot. This is best accomplished by hitting vulnerable parts with a ram or plow attached to the front of the BMW. The amount of energy contained in a straight-ahead charging robot is calculated by the equation:

$$\text{Energy}_{\text{kinetic}} = \frac{1}{2} \text{ Mass of Robot x (Velocity of Robot)}^2$$

or

$$KE = \frac{1}{2} M_{\text{robot}} \times V_{\text{robot}}^2$$

For example, imagine your 100-pound robot, traveling at 8 miles per hour, hits another robot with its ram. Your robot stops after impact and, to make things simple, we'll assume all of the energy was transferred to the other robot, which goes flying across the arena floor.

The energy transferred is:

$$KE = \frac{1}{2} \text{ Mass x Velocity}^2$$

Putting everything into consistent units of mass and speed,

$$\text{Energy Transferred} = \frac{1}{2} \times \frac{100 \text{ lb}}{32.2 \text{ ft/sec}^2} \times \left(\frac{8 \text{ mi}}{\text{hr}} \times \frac{5,280 \text{ ft}}{\text{mi}} \times \frac{1 \text{ hr}}{3,600 \text{ sec}} \right)^2 = 213 \text{ ft-lbs}$$

Thwack-bots

A thwack-bot design generally involves a two-wheel-drive design that provides it with the ability to turn like Nijinsky on overdrive. A thwack-bot turns quickly and powerfully. It is generally equipped with a long, stiff, and heavy tail. A thwack-bot uses its turning ability to whip its tail around and produce lots of angular

momentum. Like a ballet dancer executing a pirouette, the thwack-bot tries to make a solid hit with its tail and send its competitor flying across the arena. By twirling and spinning, the robot maximizes its leverage and obtains substantial rotational energy quantities.

There are several issues surrounding this type of robot. Thwack-bots are very hard to steer. The driver needs to maneuver the robot toward its opponent and then put it into a hard, sustained spin. As it spins, the robot's tail is supposed to hit into the opponent. Trouble is, the opponent is going to do everything it can to get out of the way of the tail as it spins. Most of the time, the spinning thwack-bot spends its time and battery energy doing a solo fouetté en tournant in the middle of the battle arena, while the opposing robot stands by, amusedly looking on and preserving battery power.

That is not to say that thwack-bots are bad designs. On the contrary, many winners use the two-wheel-drive and tail idea with great success. It really depends on your design, your driving ability, how fast you can crank up your spin, and how close you can get to your opponent.

The Physics of Thwack-bots

It's fun to examine the physics of thwack-bots. Take a look at a hypothetical one called Thwacky.

You want to calculate the energy developed by Thwacky when it goes into a spin. First of all, you need to know certain things about this particular robot. Its maximum speed, forward or backward, is measured about 10 MPH, or 4.5 meters/second. You also find that Thwacky's hammer weighs 4 kilograms and is mounted on a 1-meter-long handle. Twacky's overall weight, excluding the hammer, is 26 kilograms. Its shape without the hammer is basically a square 1-meter-by-1-meter box.

Thwacky's spin diameter, when the left wheel is going 4.5 meters per second forward and the right wheel is going 4.5 meters per second backward, is 1 meter. The first thing is to determine how fast the robot spins in revolutions per minute. You measure the robot's width and find that the spin circumference is 1 meter x π, or 3.14 meters. So every time Thwacky makes a full revolution, it goes 3.14 meters.

$$\frac{4.5 \text{ m}}{\text{sec}} \times \frac{1 \text{ rev}}{3.14 \text{ m}} = \frac{1.4 \text{ rev}}{\text{sec}} = \frac{86 \text{ rev}}{\text{minute}}, \text{ or about 8.6 rad/sec}$$

You need to calculate the moment of inertia, or MOI, of the robot, which in this case consists of a spinning box plus the hammer. Calculate the MOI of each component separately, and then add them together later.

The MOI of the 4-kilogram hammer mounted 1.5 meters (1-meter handle plus 0.5-meter body) from the center of rotation is found from the equation:

$$MOI_{hammer} = M \times R^2$$
$$MOI_{hammer} = 4 \text{ kg} \times (1.5 \text{ m})^2 = 9 \text{ kg-m}^2$$

The MOI of the 26-kg robot in the shape of a 1-meter-by-1-meter box from the center of rotation for the box rotating around its middle is found from the equation:

$$\text{MOI}_{\text{box}} = \frac{1}{12} \times \text{Mass} \times \text{Length}^2 \times \text{Width}^2$$

$$\text{MOI}_{\text{box}} = \frac{1}{12} \times 26 \text{ kg} \times (1 \text{ m})^2 = 2.2 \text{ kg-m}^2$$

(It's interesting to note how much more the light hammer contributes to the moment of inertia than the heavy robot body because of its distance from the center of rotation.)

The kinetic energy of a rotating system is given by the equation

$$\text{KE} = \frac{1}{2} \text{MOI} \times \omega^2$$

where ω is the rotational velocity in radians.

So the kinetic energy is:

$$\text{KE} = \frac{1}{2} \times (9 \text{ kg} \times \text{m}^2 + 2.2 \text{ kg} \times \text{m}^2) \times (8.6 \text{ rad/sec})^2 = 414 \text{ joules}$$

So how hard will Thwacky hit? Remember that this is a lightweight robot and with motors and batteries tips the scales at about 30 kilograms. If Thwacky's hammer struck another 30-kilogram robot from underneath, it would convert the kinetic energy of the hammer to potential energy by raising the other robot off the ground to some height, before gravity brings it back down. The height to which it is raised is figured from the potential energy equation (the gravitation constant, g_c, in metric units is 9.8 meters/second^2):

$$\text{PE (joules)} = \text{Mass (kg)} \times \text{Height (meters)} \times 9.8 \text{ meters/sec}^2$$

So the height = 414 / (30 x 9.8) = 1.4 meters.

Therefore, 414 joules is enough to smack a 30-kg robot to a height of 1.4 meters into the air.

By contrast, if Thwacky decided to simply charge its 30-kilogram body into the opposing robot, at its maximum speed of 4.5 meters per second, it develops:

$$\text{KE} = \frac{1}{2} \times 30 \text{ kg} \times (4.5 \text{ m/sec})^2 = 303 \text{ joules}$$

This is enough energy to raise the opposing robot

$$\frac{303}{(30 \times 9.8)} = 1.0 \text{ meters}$$

or about 30 percent less than a maximum speed spinning hit. The extra energy you can obtain from good thwack-bot designs accounts for their popularity.

There is a subset of thwack-bots that incorporate gyroscopic or microprocessor control. These recent additions to the fighting robot scene utilize onboard intelligence that allows the robot to rotate and move toward an opponent at the same time. This is a difficult engineering problem indeed, requiring the use of sophisticated sensors.

Cutting Blade Robots

Cutting blade robots employ large hardened cutting wheels or saw blades to cut through the armor of competing robots. The cutting

14.3 The Kapitan. COURTESY TEAM KARL

blades are generally driven by belts, gears, or chains. Driving the saw blades are gasoline or electric motors, as large as weight and size constraints will allow.

A single good solid hit by a powerful cutting blade robot can incapacitate or even decapitate a robot. Getting a solid hit can be difficult, however. The saw blades tend to bounce off armor, and getting good purchase to start the initial cut can be problematic. Large teeth seem to do much better—they rip and toss or flip. Small teeth won't do much in these applications.

Judges and spectators seem to love cutting blade robots. Maybe it's the noise, the motion, or the undisguised menace of spinning teeth, but win, lose, or draw, they're always popular.

Lifters

Ah, lifters. These are among the most exciting and successful of the many types of robot weapons. A lifter robot has mechanical arms, powered by either electrical motors or high-pressure compressed fluid cylinders, that are designed to get underneath the opposing robot. Then, in the blink of an eye and with a single great effort, they lift the opposing robot quickly, sending it ass-over-teakettle across the arena floor. Few robots are built solidly enough to withstand being flipped more than once or twice. See Chapter 15 for information on designing fluid power flippers.

Chameleon Robots

The A-20 Warthog fighter-bomber was the star of the Gulf War in Kuwait. It wasn't the fastest plane in the sky, nor the biggest, but it was the most versatile. It could be outfitted with missiles, cannons, smart bombs, dumb bombs, bombs of average intelligence, bags of rotting vegetables, and so on, depending on the orders received from the Pentagon that day. Some clever robot builders follow that same Swiss army knife philosophy and build flexibility into their robot in the form of rapidly attachable and detachable weaponry.

For example, a robot going up against a spinner may attach a long, thwack-bot-like appendage, designed to whack the spinner without bringing the main robot body into close proximity with the spinner. That same robot may be outfitted with a hardened steel spike to penetrate the wheels or armor of a BMW. It may also allow the attachment of low-clearance armor skirts to its outside frame to prevent a lifter from getting underneath it to give it a flip.

Like a chameleon, a robot with flexible weaponry is readily adaptable to a variety of situations. It brings the whole panoply of robotic armaments to bear, and allows the builder to exploit weaknesses in its opponent's design.

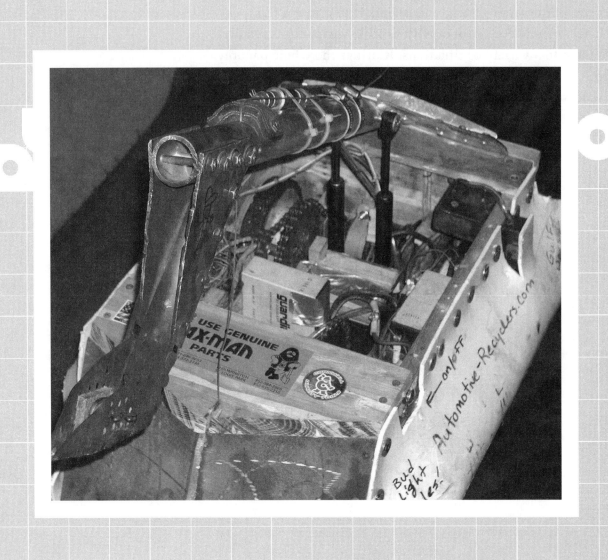

15.
FLUID-POWERED WEAPONRY:
FLIPPERS, GRIPPERS, AND CLAMPS

Some of the most successful and admired fighting robots use fluid power to operate their weapon systems. Fluid power involves the use of pressurized gas (pneumatics) or liquids (hydraulics) to provide a mechanical movement. Fighting robots utilize fluid power to manipulate and control flipping arms, shovels, pincers, movable spikes, and lots of other nasty stuff. This chapter is an introduction to this subject, which is among the most complex in this book. It takes considerable building and refining skill to fabricate an effective weapon using this technology. The information here can at least get a builder started in the right direction, and a good builder may be able to take this kernel of information and make a very powerful fluid-powered weapon for his or her bot.

Before you run out and start experimenting with high-pressure fluids, understand this: fluid power is well tested, versatile, and powerful, but in the wrong hands *it can be very dangerous.* Here are some important safety guidelines for building pneumatic and hydraulic robot weapons.

Fluid-Powered Weapon Safety

The following safety rules are just for starters. Working with fluid power requires caution. Just imagine a high-pressure hose coming loose and whipping around at lightning speeds like an angry snake!

1. All of the previously discussed safety rules apply, but even more stringently.
2. After reading this, you will have just enough knowledge to get started. Know what you're doing or ask for help from someone who does. If you don't know what something means, ask someone.
3. Wear safety goggles and gloves when working with pressurized systems.
4. Limit pressures to a safe maximum. Start with very low pressures and work up slowly. Know exactly what your equipment is rated for and do not exceed those limits.
5. Always use pressure bottles/vessels/reservoirs of a commercially acquired type. This means they must be specifically manufactured to contain gas at pressure and have a government-approved rating. Of course, the pressure you use must never exceed the manufacturer's specifications.
6. Pneumatic and hydraulic lines and fittings must be built in accordance with government-sanctioned national standards and specifications.
7. Route all tubing, piping, and hoses so as to minimize the chances of being cut or damaged during a match.
8. All fluids in fluid power systems must be inert, or at least nonflammable.
9. Locate gas bottles, reservoirs, valves, and regulators in your robot's interior to protect them from puncture or other serious damage.
10. Gas bottles and reservoirs must be securely fastened down, and the valve/regulator must be strapped or otherwise held securely.
11. Your robot should have a pressure relief safety valve fitted on the high-pressure side of the regulator.
12. You should be able to vent any part of the pneumatic system using a valve. If you have to take a wrench and remove fittings to remove pressure, that won't do!

13. Use gauges on your system so you know what the pressure is at both the high- and low-pressure parts of your system.
14. Make sure that your system conforms to the most stringent rules and requirements. Event safety marshals are very picky when it comes to pneumatic and hydraulic systems.

What Is Fluid Power?

When you transfer energy from one component to another using pressurized liquids or gases, you're utilizing fluid power. Pneumatics uses gases to transfer the energy. Typical gases used in fighting robots are compressed air, carbon dioxide, and inert gas. Hydraulic systems use liquids, which are normally specially designed oils.

The big lifting arms and hammer weapon systems using fluid power work because forces are transferred, via the fluid, from a high-pressure reservoir to a movable surface, such as a piston inside a cylinder. The surface (e.g., piston) moves forward when the force pushing the piston is larger than the total load pushing back, plus any frictional forces. So if you design a lifter arm with a big enough piston and cylinder, and with high enough pressure acting on it, you can move an awful lot of weight in a hurry.

Fluid Power Advantages

Using fluid power has some advantages over using, say, electrical actuators and motors. First, you only need a single source or reservoir of high-pressure fluid. A single fluid pressure storage device (e.g., a tank) can power many motion devices. More important, the power source can be located where space isn't critical; you can place your large pressure tanks wherever you have space for them.

The energy in the system comes from fluid compressed by big machines that are *not* onboard your robot. The compressors are in the pit area, not on your robot. Therefore, the system's size and weight is small compared to all the energy it contains. You get a lot of power and only need to use some fairly light onboard components—the cylinders and actuating equipment. These types of weapons can be quite small and light and still have immense power.

In weapon systems using big jaws or clamps to clamp down on opponents, pneumatic or hydraulic grippers can produce a strong and constant holding pressure. Since the actuators use no energy while they're stationary, that's a big advantage over electric motors. Motors draw considerable current to maintain torque, even while stopped. Indeed, many motors will overheat and fail under these conditions.

Hydraulic Versus Pneumatic Power

Robot builders have used both hydraulic and pneumatic power systems successfully. Indeed, both share many common characteristics, although there are some key differences. Pneumatic systems are typically easier and less expensive to build. Hydraulic power is more precise because the liquid doesn't compress. Also, tournament rules sometimes allow them to operate at higher pressures.

Designing a Pneumatically Powered Flipper Weapon

Pneumatics Basics

The gas used in pneumatically operated robots is high-pressure air (often referred to as HPA), nitrogen, or carbon dioxide. HPA and compressed nitrogen have similar properties. HPA is stored as compressed gas in a strong pressure tank at pressures of 1,000 psi and higher. The gas is metered out of this storage bottle by means of a regulator, which fits atop the storage tank. The output of the regulator is normally 100 to 200 psi.

Carbon dioxide works similarly, except that it is stored in tanks in liquid form. When it is regulated down to lower pressure, it undergoes a phase change and becomes gaseous. As it does so, it absorbs heat from its surroundings, so it and the equipment get cold, sometimes very cold. The cold causes different types of frosting and freezing problems, so extra care must be taken when you design with CO_2. The use of accumulator tanks (explained below) and anti-siphon tubes is highly recommended.

So why bother to work with CO_2 at all? Because CO_2 components are generally smaller, lighter, and less expensive than HPA. Many of the CO_2 parts—regulators, storage tanks, and tubing—are readily available at stores that sell paintball equipment.

Fluid Power Components

Designing the fluid power circuit, as well as selecting and physically placing system components, is critical to maximizing the robot's weapon performance. The diagram below shows a typical fluid-powered weapon using pressurized air for its power source. This pneumatic system is made up of a supply tank, an accumulator tank, a valve, a solenoid valve operator, a cylinder, and various and sundry assorted tubes, connectors, and fittings. Each part is explained in the sections that follow.

Pumps and Tanks

Pumps supply the air or gas under pressure to move the system's cylinder pistons, converting mechanical power to fluid power. Usually there is no pump or compressor on a robot. Instead, the builder will pressurize a large holding tank that resides onboard the robot. The pump is external and is used only to pressurize the holding tank to a safe pressure (which depends on the cylinder; each is rated by its manufacturer). Once the tank is empty, the weapon will no longer work. Therefore, the tank must be sized carefully. It should be large in terms of volume and pressure to store enough gas to actuate the cylinder as many times as needed during the match.

If you size it too small, you will run out of air pressure too soon and your weapon won't work at the end of the match. Size it too large and you'll be dragging around extra weight. Sizing the cylinder requires you to do careful calculations beyond the scope of this chapter. Most builders choose a convenient cylinder size and safe pressure and then figure out how many times their particular arm can be used. Then they simply compete with that number of "flips" in mind.

The pressure of either CO_2 or HPA inside a storage tank can be upward of 1,000 psi. The pressure must be stepped down to a lower pressure that is compatible with the other weapon system components. This is accomplished by installing a regulator on the tank, which meters out CO_2 or HPA at a lower pressure, typically 100 to 200 psi.

If you use CO_2, you should outfit your tank with an anti-siphon tube. This device is designed to prevent liquid CO_2 from entering the pneumatic system. It is an angled tube permanently connected to the tank's valve and positioned so that the tube opening is immersed in gaseous, not liquid, CO_2. Doing so prevents the introduction of liquid CO_2 into the regulator and beyond. Paintball equipment suppliers are excellent sources of pressure tanks, tank valves, anti-siphon tubes, and pressure regulators.

Cylinders

Cylinders convert fluid power to linear mechanical power. Cylinders can be single-acting or double-acting and are rated as to the maximum pressure they can accept. A single-acting cylinder is powered by air in one direction (extend) and is usually returned to its retracted position by an internal spring. A double-acting cylinder is powered by air in both directions. These are available as two-position (extended or retracted) or three-position (extend, retract, or off).

For either type, pressurized fluid enters one side of the cylinder and pushes against a piston. The faster the fluid enters and the larger the piston's surface area, the faster and harder the actuator works. As with tanks, correctly sized cylinders are important. Increasing a cylinder's diameter provides more force for a given level of pressure, but also requires larger tanks, valve diameters, and tubing.

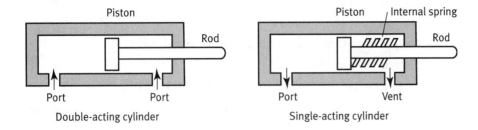

Double-acting cylinder Single-acting cylinder **15.1 Cylinders**

Piston *Rod* *Port* *Port* *Piston* *Internal spring* *Rod* *Port* *Vent*

Pressure Regulator

A pressure regulator is a metering device that separates the uncontrolled high-pressure gas in the supply tank from the controlled gas pressure in the lower-pressure parts of the system. Often the collection of parts of the fluid power system on the upstream side of the regulator is termed the high-pressure side, and the collection on the downstream side is called the low-pressure side. A pressure regulator is designed to take whatever pressure is in the supply tank and output a user-specified pressure to the cylinders and other moving parts of the system.

Accumulator Tanks and Buffers

Accumulators are medium-pressure fluid storage devices located downstream of the main pressure regulator. Accumulators serve an important function. They store large volumes of air at the regulated pressure, so when you open a cylinder valve, you can get a lot of air in a hurry without being bottlenecked by the narrow diameter opening of the pressure regulator.

Hoses and Tubes

Tubing, piping, and hoses are used to direct the fluid between components. In general, keep them as short and straight as possible because pressure drops occur in bends. Try to locate valves as close to the cylinders as possible.

Valves and Valve Operators

Valves control the entry and exit of fluid to the cylinders. Many builders say that the valves are the most important part of the system; it seems that good, reasonably priced, high-flow-rate valves can be hard to find. They are often operated electrically by a device called a solenoid. The solenoid is controlled by either a servo motor or an electrical relay, either of which is controlled using one of the channels on the R/C transmitter system.

Solenoid valves may have special "air assist" features to make them operate faster. Such features are called pilot valves, and they are often needed to create the high-speed movement of the lifting arm.

Pneumatic valves are designated in terms of the number of ports they have and the number of ways the ports are connected. Most manufacturers of fluid power valves offer a wide variety of ports and ways. The most important valve types in robot building are the two-way, two-port valve used for controlling single-acting cylinders; and the four-way five-port valve used to control double-acting cylinders.

Design Examples

The best way to understand pneumatic weapon design is to consider a couple of examples. To design a fluid-powered robot

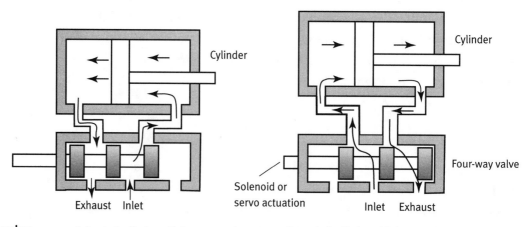

15.2 Four-way spool valve   Retracted cylinder with four-way valve   Extended cylinder with four-way valve

weapon, start with an end in mind. Assume that after looking at many different possibilities and designs, we decide to build a lifting arm robot. Examine the lifting arm example shown in Diagram 15.1. In position one, the arm is down and the tines are parallel to the floor. In position two, the arm is up and the tines have lifted the opposing robot into the air, and hopefully tipped it over.

In order to obtain the lifting action, an actuating arm pushes against a lever, hinged at one end. Since the hinged end can't move, the actuating arm imparts a rotary motion to the tine end, and this rotary motion scoops the opponent up and over.

In order to design a hinged lifting arm, you can use a rigorous analytical approach using trigonometry and computer design software, or you can use the T-LAR method. There is no doubt that the better approach is the former, but that is beyond the scope of this book. So we'll go with the "that looks about right" (T-LAR) method and use trial and error to refine the arm.

15.3 Designing a flipper arm

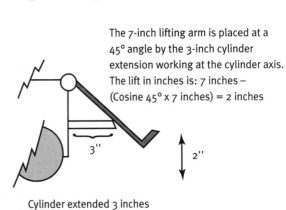

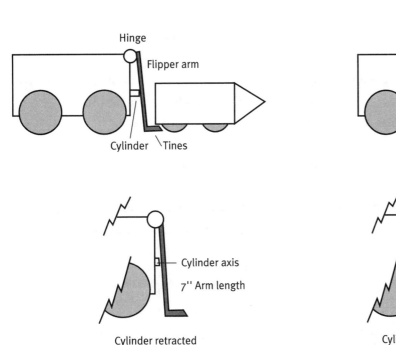

The 7-inch lifting arm is placed at a 45° angle by the 3-inch cylinder extension working at the cylinder axis. The lift in inches is: 7 inches − (Cosine 45° x 7 inches) = 2 inches

Cylinder retracted

Cylinder extended 3 inches

The amount of lift can be increased either by moving the cylinder axis up toward the top, or by increasing the amount of cylinder extension.

Here are the design criteria for a robot called "Little Lifter":

- You need the cylinder to extend a total length of 3 inches.
- You need the arm to have enough power to push with 200 pounds of force.
- You need the arm to raise fast enough to impart a flipping motion to the opponent.
- You want to use pressurized air as the power source.

Based on your design criteria, determine the size of the cylinder you need to obtain. First, determine how much air pressure you can use at maximum. A glance at the tournament rules will normally provide this information. For the tournament that Little Lifter will enter, the rule book tells us that 150 psi is the maximum allowable pressure that can leave the high-pressure storage tank regulator and go into the cylinder.

The following equation shows how to determine the force applied by a cylinder at a given pressure:

$$\text{Force} = \text{Pressure} / \text{Area}$$

Your design calls for pushing with 200 pounds of force, and you also know that the pressure in the cylinder is limited to 150 psi. Therefore, the required area of the cylinder is:

$$200 \text{ lbs}/150 \text{ psi} = 1.33 \text{ square inches}$$

Since the piston is round, you know that the area of the piston must be $2 \times \pi \times \text{radius}^2$. Solving for the radius required is:

$$\text{Radius} = \sqrt{\frac{1.33}{2 \times \pi}} = \text{about 0.46 inches.}$$

Therefore, you need a cylinder with about a 1-inch diameter and an actuation range of 3 inches.

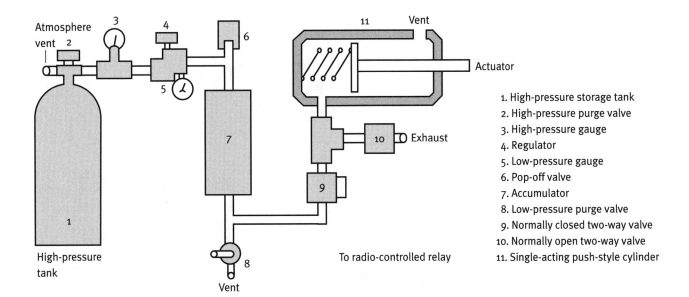

Atmosphere vent

High-pressure tank

Vent

Vent

Actuator

Exhaust

To radio-controlled relay

1. High-pressure storage tank
2. High-pressure purge valve
3. High-pressure gauge
4. Regulator
5. Low-pressure gauge
6. Pop-off valve
7. Accumulator
8. Low-pressure purge valve
9. Normally closed two-way valve
10. Normally open two-way valve
11. Single-acting push-style cylinder

A simple lifting arm system is shown in Diagram 15.4. This design makes use of two simple valves and a single-acting cylinder with a spring return. Here, a solenoid-operated valve is actuated and high-pressure gas is directed against a spring-loaded piston inside the cylinder. If everything is sized right, the gas enters quickly and forcefully enough to flip whatever is underneath the actuator. Then, the supply valve closes and the exhaust valve opens, venting the cylinder back to the atmosphere. The spring returns the piston to the starting position, and the system recharges the accumulator for the next flip.

15.4 Single-acting cylinder flipper system

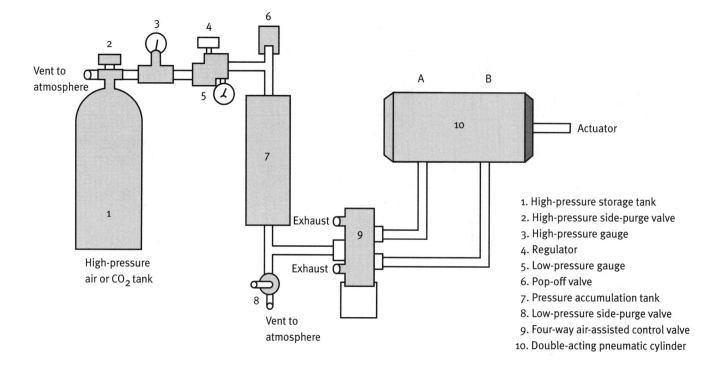

Vent to atmosphere

High-pressure air or CO₂ tank

A   B

Actuator

Exhaust

Exhaust

Vent to atmosphere

1. High-pressure storage tank
2. High-pressure side-purge valve
3. High-pressure gauge
4. Regulator
5. Low-pressure gauge
6. Pop-off valve
7. Pressure accumulation tank
8. Low-pressure side-purge valve
9. Four-way air-assisted control valve
10. Double-acting pneumatic cylinder

15.5 Double-acting cylinder flipper system

You can accomplish a slightly more elegant design using a five-way valve as shown in Diagram 15.5. This design allows the cylinder to extend and retract more quickly because high-pressure gas, instead of a spring, controls all cylinder movement. The lifting device will follow this course of events in order to make a lift.

1. The operator presses the lift button on the radio controller.
2. An R/C servo motor or R/C relay shifts the four-way valve to the "extend actuator" position.
3. When activated, high-pressure gas from a reservoir rushes through the valve into a cylinder and pushes the piston out.
4. When the R/C signal for "retract actuator" is received, the gas is released from the A side of the cylinder and vented to the atmosphere. Simultaneously, gas is directed to the B side of the cylinder, which causes it to retract to its starting position.

One problem that builders frequently encounter is that of restricted airflow into or out of the cylinder. In this case, the cylinder has plenty of power and plenty of volume, but it's just too slow to extend. Try some of the following design tips to increase actuator speed:

- Use a bigger buffer or accumulator tank.
- Use the biggest pipe and tube fittings that you can.
- Use large diameter hose and tubing.
- Avoid kinks, turns, and sharp corners in the piping and tubing runs.
- Make the gas exhaust path out of the cylinder as free from restrictions as possible.

16.
GAME DAY STRATEGY
AND TACTICS

As any winning robot builder will tell you, strategy, tactics, and driving skill play a major role in the outcome of most matches. To be fully prepared, you need to plan your tactics well before your bot is placed in the fighting arena.

Before the Match

Take a few minutes and carefully evaluate both your robot and the one you're matched against. Determine what advantages you have and what advantages your opponent has. There are a number of important things to consider:

- **Who is faster?** Many builders disclose performance details such as the robot's top speed to just about anybody who asks. If your opponent won't, look at the motors and the drivetrain if you can see them, for a clue.
- **Who has more torque?** Again, take a look at the drive motors and the gear or chain sprocket ratios that connect the motor shaft to the wheels.
- **Where are the weak spots in both your opponent's and your own armor?** Are the wheels pneumatic,

Pit Area

HOW TO PREPARE FOR YOUR FIRST MATCH

From Stephen Felk, builder of Voltronic

Believe it or not, a lot of fights are lost due to mental errors—being too tired or too nervous. Try and get the bot done early. (Yeah, right!) Get some time off before the event so you are well rested before it starts. Try and get as much driving practice in as possible. Make sure that you take the bot out of the house and drive it at least a couple of times before the match. Load the bot and your gear and go out to some public place that is safe and get some driving practice. It helps mentally to deal with running the bot away from the comforts of home. It also helps to be out in public with the bot where you have to deal with onlookers.

Overcoming nervousness is all about feeling comfortable. Basically, you are going to be nervous; everybody is, especially during the first match of the event. You can learn to expect and eventually love this feeling. Think of adrenaline as "focus juice." Find out what makes you comfortable. It may be as simple as wearing your favorite shoes or brushing your teeth before your match. Warm-up stretching is good. Some builders visualize the match while they are waiting, while others talk like banshees before they fight as a way of thinking about something else. And the best thing is having lots and lots of driving practice.

solid rubber, or foam filled? Are they protected by Lexan or steel skirts, or is there room to make a run on them with a spike?

Determine your strengths against your opponent's weaknesses as best you can and then formulate a strategy to maximize those.

Specific Strategies

Spiked Robots

A robot bristling with long, sharp spikes can be tough on its rival's wheels and armor. But—and this is a big but—robots with fixed spikes often become immobilized because they get impaled on the arena walls. Many an otherwise undamaged robot has lost a match because it has been rendered immobile by allowing a sharp appendage to become stuck deep into a wall while its adversary sat and watched the clock run down. If you have spikes, stay in the middle of the arena. If you're fighting a spiked robot, stay near the walls.

Spinners

If the robot you are facing has a heavy spinner weapon, it must be designed stoutly enough to handle that first big hit. This being said, it may be possible to avoid a full-speed impact with the spinning weapon. First, figure out how long it takes to spin up to full speed. If it takes say, eight seconds to spin up, that gives you a small but significant window of opportunity to try to hit it before it is at full

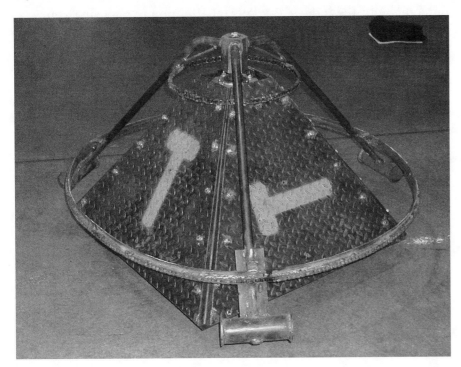

Thor. COURTESY TEAM NORSE

speed. Second, always attack a horizontally spinning robot by going into the direction of his spinner. Even better, hit it when it is close to the arena wall, so the energy in the spinner is dissipated into the wall. Third, spinners tend to be really slow, since most of their weight is directed into the weapon-spinning motor and the kinetic energy shell, and not the drive motors. This can be used to your advantage. If your robot is much faster, you can outmaneuver a spinner and force it into arena hazards and walls.

If you need to engage after the spinner is at full speed, try to hit the opponent into an arena hazard or into a wall to take some of the energy out of the spinning weapon.

BMWs

If your opponent is a BMW, determine which side or end is most vulnerable to your own weapons. These guys tend to be powerful, fast, and heavily armored since no weight is tied up in weaponry.

Judges tend to dislike BMWs because they are not considered very artistic or exotic, so if this is the type of robot you build, inflict as much damage with your weapons as possible for a positive outcome with the judges. Pay special attention to choosing your wheels. Choose them to maximize traction, since this type of bot should be optimized to push other bots.

Stinger. COURTESY A. J. KLEIN OSOWSKI

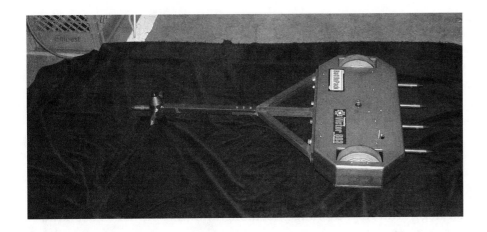

Baby Gouda. COURTESY TEAM CHEESEHEAD

Thwack-bots

If your opponent is a thwack-bot, carefully note the radius of the attacking appendage so you can stay out of range during its spin cycle. Patience is the key in fighting a thwacker. At some point, the thwack-bot's driver will tire of spinning around mindlessly, exhausting its battery, while you sit there watching it. The thwack-bot will stop its spin in order to maneuver. That's the time to strike.

Lifters

If the robot you're facing has a lifting arm, try to determine its weight capacity and estimate the number of lifts it is capable of

Toro. COURTESY BATTLEBOTS, INC., AND INERTIA LABS

making based on the size of its CO_2 or HPA chamber. Obviously, avoid letting it get its lifting arms underneath your robot.

Game Plans

The first five seconds of a match can very often determine the eventual winner. For example, let's say your robot is a BMW and you are up against a powerful spinner. This calls for a "joust strategy" where you plunge headlong into the spinner as soon the match begins, hopefully before the spinner's motors can bring the shell up to its full speed.

If you are driving the spinner, conversely, you should drive evasively to avoid full contact until the spinner is at maximum RPM.

After the initial joust, if there is one, remember that the judges award points for aggressiveness and your goal is to get as many good hits as possible on your opponent.

Inflicting or sustaining visible damage is very influential in the judge's decision. The estimation of damage is very subjective. What looks to be a massive injury may not affect the robot's performance much, but because it looks so bad, the judges award tons of points. Along those same lines, visual and aural indications of weapon effectiveness are very influential. The roar of a saw blade attempting to cut steel armor often sways judges. Sparks from a carbide-tipped saw digging against tool steel can also convince judges to award damage points. Therefore, make your weapon loud and sparky if you can! When you're looking for targets, consider your opponent's wheels. Many robots have exposed wheels that make pretty good targets. Take a run at the wheels if you think you can do some damage there.

Very frequently, one or both robot's primary weapons become disabled during the match. For all intents and purposes, you are now a battery, motor, and wheel type of robot, and you now need to decide whether to pursue a power or a momentum end-game strategy.

If your robot can push harder, go for a power approach. Engage in close-quarter fighting and push your opponent into the ring hazards if there are any. If there are no hazards, push it into the wall. Just push, push, and push again.

If your robot is designed for speed, not power, then you should adopt a momentum strategy. Momentum is a Newtonian physics quantity that is the product of mass and velocity. The faster you go, the more momentum you develop. You can think of momentum as your ability to wallop someone. Therefore, if you have a speed advantage, take long, high-speed, cross-court runs at the other robot and aim for its most vulnerable parts, like wheels.

If your weapon becomes inoperative and your opponent can push harder and is faster than you to boot, you're pretty much screwed. At this point, try to avoid getting damaged (to the extent you can) without looking like a complete coward. You can try to lure your opponent into an arena hazard by playing rope-a-dope with him, but your chances of success aren't great.

One Final Word

Practice makes perfect. The best competitors know that a skilled driver can make an average bot into a great one. There is just no substitute for driving practice. Turns, reverses, spins, wheelies, whatever your robot can do, you should practice to such an extent that it becomes second nature. Experience driving an R/C car doesn't really count; cars and robots handle entirely differently. The best drivers can control their bots like an extension of themselves. That can only come from a great deal of practice driving.

10 Tips for Building Battle-Worthy Robots

1. Many inexperienced builders use setscrews to affix sprockets and bearings to shafts. Unless you have an extremely lightweight robot, this is generally a mistake. They are not strong enough to handle the torsional forces your robot will experience. All of a sudden, your motors will whine and spin up, but the wheels or

weapon will just sit there, idly. Yup, another setscrew bites the dust. As one well-known robot builder succinctly puts it, "Setscrews suck." Instead of setscrews, broach your shafts at a machine shop or buy pre-slotted shafts and use keystock to secure these items in place. If this isn't practical, then pin your shafts to the sprockets by drilling a hole through the shaft and then use a machine screw and nut to keep it secure.

2. Your safety cutoff switch should be designed such that it is easy to get to when you need to turn the machine on and off. But it should be mounted in such a way that it won't get accidentally switched off by impact or a stray hit from your own weapons.

3. The connection to your batteries is supremely important. Lose the connection to either battery and the match is lost. There are two main types of connectors: spade connectors and screw down connectors. Screw down connectors are more secure than spade connectors. If you are going to use spade connectors anyway, solder or tape them down securely.

4. Don't run out of juice at the end of a match, just when you need it the most! Determine the number of ampere-hours required for your match as described in the section on batteries. Then go out and buy the appropriate type of batteries that are big enough and include a margin of safety. How much margin? It depends on how much weight you can give up. Also, along those same lines of thinking, make sure your batteries are fully charged before each match.

5. Make absolutely certain that your speed controller, internal wiring, cutoff switches, terminal strips, et cetera, are rated to handle the current loads that your robot will generate during the match. Not only will you lose the match if your speed controller smokes, but speed controllers are expensive.

6. The antenna is a very important but often overlooked component. Make certain it is long enough, and mounted in such a fashion that it can receive your transmission clearly from any part of the arena.

7. The radio receiver, your batteries, and other parts are hard to mount because they don't come with bosses or brackets for mounting to the frame. Just the same, they must be mounted so

that if your robot gets rocked, socked, or flipped, the connections will still be secure and tight. Use special mounts, U-bolts, or clamps. Although I don't recommend them in larger robots or for parts that get hot (like batteries), multiple tie-wraps, which are ratcheting plastic ties that hold items in place, can work satisfactorily.

8. A chain is only as strong as its weakest link. If your power transmission train utilizes sprockets and drive chains, be sure to check the master links to make certain the spring clip is securely in place.

9. If your opponent is savvy or experienced, there is a high likelihood he or she will go after your tires. Protect your tires with armor, make them of solid material, or fill them with foam so they can run flat.

10. Finally, make sure all mechanical and electrical connections are as secure as possible. Use electrical ring connectors instead of spade connectors, and solder the connections where appropriate. You can wrap mating male and female connectors with electrical tape if you think that will help. Bolted connections tend to work loose during a match. Therefore, consider using chemical threadlockers such as Loctite where appropriate.

17.
THE ENTREPRENEUR'S GUIDE TO BATTLING BOTS

This chapter covers a nontechnical aspect of fighting robots—the business end of things. When you add up everything it takes to compete in this sport, it is obvious that robots cost money, sometimes lots of money. What is less obvious is that robots can earn money as well. A builder who understands the value of his or her creation and, further, can draw from the experience of business-savvy automobile racers, monster truck builders, and event promoters can turn this hobby into a profitable sideline, or at least a less costly endeavor.

Getting a Sponsor

Here's how many people approach the idea of sponsorship. They sketch up a design for a robot and then figure out how much the beast is going to cost in terms of parts and equipment. When the total exceeds the amount they've got to spend, someone in the group says "Hey, let's get a sponsor to pay for this or to give us free parts." There's no way anyone with a shred of business sense is going to give an unknown robot builder a dime to build a titanium monkey-like vapor-bot with a prehensile flair tail. It's just not going to happen.

Ricky Ticky COURTESY TEAM
NEVERMIND

In order to get a sponsor, you have to put yourself in the sponsor's perspective. The key concept is to think like a business owner. Say you own Joe's R/C Car Shop. From a business standpoint, think about why Joe would want to sponsor your robot. Before you say "advertising value," think how much radio and print advertising that store could buy for the money it would spend sponsoring your robot. Unlike an ad on the radio or in a trade magazine, just plastering the name of the store on the side of your robot will not make people come in and buy R/C car parts.

The fact is, placing a sponsor logo or name on the side of your robot will not justify the cost of sponsorship. Even if you make a television round at a large national tournament, a quick flash of the sponsor's name on the side of the robot won't make a large impact in your sponsor's sales.

When it comes to sponsorships, think "relationship." Think of ways to initiate a relationship that could lead to a sponsorship. There are lots of simple, no-cost things a builder can do to get a relationship started. Be willing to put a potential sponsor's logo or stickers on your new robot without any money in return, at least in the beginning. Think how a potential sponsor would feel if he saw his logo on your robot during a match or television program. Would the sponsor appreciate it? You bet!

Chat up people who might be interested in working with you. Be friendly and businesslike, but never desperate. Offer to let your potential sponsor be in your pit crew, and hang out with the other builders in the pit area. Invite the potential sponsor to your workshop to drive the robot and have some fun with it.

Buy the product or service the potential sponsor offers. If it's good, tell other people. And it never hurts to make sure the potential sponsor knows how you're helping generate business.

Think Partnership with Your Sponsor

Once you have a relationship, then you can pitch your plan for a partnership with your sponsor. In the NASCAR world, the term "unique visibility" is used to justify why people and companies provide money. The stockcars painted like Tide detergent boxes and Viagra bottles are examples of this. (Think about this: a robot with a big pneumatic lifter arm, sponsored by Viagra. Wow!) Take the idea of unique visibility and go further with it. Propose using the robot to attract attention at a store sale. Offer to drive the bot at parades and civic events. In the motor racing world, racers often paint their personal street car or truck exactly the same (down to the numbers and stickers) as the sponsored race car. That may be going too far, but you get the idea. Also, this requires that your robot live up to the sponsor's image. For example, Joe's R/C Car Shop would not want some beat-up junk pile representing its business. The robot needs to look good, and so must be repaired promptly after each tournament.

Think of ways you can fully use the power of your robot's unique visibility. This power can be used at more than the tournament. It can be used at the sponsor's business or away from the business in places like shopping centers or intersections. Any place that has good traffic is a target for your billboard. Now go the extra step—have a professionally painted sign saying, "Come into Joe's R/C Car Shop and get discount tickets to see the next robotic combat tournament." Or, "Come into Joe's Shop and sign up for a chance to drive the robot," or any other idea that will cause people to go to Joe's business. The robot is an eyecatcher, so use it to convey a message.

In summary, the two most important concepts in obtaining sponsorship is to think relationship first and partnership second. First, prove yourself. Then, build a promotional package with as many opportunities for your sponsor as possible.

How to Organize Your Own Fighting Robot Event

Currently, there are several dozen fighting robot tournaments held around the country. Some are television driven, with big budgets and celebrities, and some are gritty, garage venue affairs held just for the amusement of the builders. Given the tremendous interest in the activity, there is little doubt that more events, big and small, will start. Actually, there are many areas in the country that have a large contingent of current and future builders and are just waiting for a promoter to fill that need by organizing a fighting robot event.

Do you have the vision and the drive to be the next successful fighting robot event promoter? Somebody will do it; it might as well be you!

Before you bite off more than you can chew, be aware that organizing a tournament is complex and time-consuming. But undoubtedly it can be done and many organizers even (well, sometimes) make money as well.

If done right, a fighting robot tournament event can be a source of income for you or even a group of like-minded individuals.

Here are some important guidelines from tournament organizers who have put together similar events successfully:

If you can work with people, you can do it! Planning and organizing an event such as this takes a lot of energy and people skills. The first consideration is to get enough support and people to participate in the event, and enough workers and volunteers to organize and run the event. Your builder friends, family, and coworkers are a good place to start. It takes people—and depending on the size of the tourney, a lot of people—to have a good tournament.

This is definitely a group project. Therefore, you will need to adhere to some basic rules about working within groups.

- First, people will only be comfortable working with you if they believe you know what you're doing. Have a preliminary plan worked out prior to the first meeting with your group. People will react far better to you if you are organized and prepared.
- Second, recognize that people are good at different things. It will be fruitless to ask an introverted engineer to try and sell sponsorships to local companies. But he or she may excel at designing the arena. Utilize your team's talents and search out people with the skill sets you need and don't have.
- Third, realize that organizing an event is more about people than machines. You will have to coax, cajole, convince, and constantly remind. Be prepared for a crash course in applied human relations.

Set Up Your Time Line

Give yourself plenty of time to get things ready. You'll need time to plan and organize when, where, and who will be there. You'll also need time to establish relationships with sponsors, venue managers, and vendors of services such as the building, insurance people, and so forth.

Put together a plan that outlines the time line of the event. Begin by identifying the key dates. The most important date is obviously the date of the tournament. Start with this date and work backward.

When choosing the tournament date, think carefully about conflicting events that may interfere with participation from prospective builders. You may want to avoid times like holiday weekends or college finals weeks if they would adversely affect builder turnout. Likewise, if you plan to invite the public and sell tickets, you might avoid days with competing events such as the Super Bowl, a local monster truck show, and so forth. Once the event date is chosen, work backward from that date and choose the other key milestone dates. Milestone dates might include items such as:

- **Builder Registration Begins.** This is the date on which your organization begins to accept registrations from robot builders. Dates and times of the events should be locked in, or at least pretty well finalized. The rules of the event should be complete and published. In short, the information that builders will need to begin to build their bots must be available.
- **Builder Registration Deadline.** In order to keep the event orderly, a cutoff date should be established after which no further builders will be entered. In order to aid planning, many events offer a discount for early registrations, for example, those obtained four months before the event, or an additional fee for late registrations.
- **Venue Rental Agreement Finalization.** Finding a good venue can be difficult. Locking in the appropriate venue is a major milestone.
- **Insurance Coverage Finalization.** Insurance is vital. Do not wait until the last minute to obtain insurance.

- **Pre-tournament Builder Meeting.** A series of informal builder meetings, held every month or two weeks in the period leading up to the tournament, will help ensure clear expectations among builders and the event's managers. Such meetings may be simple affairs held in local coffeehouses, or more hands-on sessions held in workshops.

How to Manage Staff and Volunteers

First, determine the required job functions and then enlist help from people with skills that match your requirements. Here are some of the skill sets that will likely be required:

- **Technical Committee Members.** They will define the event rules (often done by a "Rules Subcommittee"). The rules committee decides the standards to which robots will be built, what weapons are allowed, what the weight classes are, the length of the matches, and the judging criteria. An example of tournament rules is provided in Appendix III.
- **Event Marshals.** On tournament day, these folks carefully inspect each robot for conformance with the rules and weigh it to determine the correct weight class. They escort the robots into the arena and make sure that all safety and other procedures are followed.
- **Builder Relations.** This person runs the builder information sessions, assigns radio frequencies, and answers builder questions. A very important job!
- **Ticket Sellers.** Someone must sell tickets to the public, if this is to be a public event.
- **Venue Managers.** They obtain a suitable venue for the event, design and build the arena, obtain insurance and permits if necessary, and handle other site-related needs.
- **Stage Crew.** The stage crew erects the arena, designs and builds arena hazards, and organizes the pit areas, press areas, bleachers, and so forth.
- **Media Relations.** This person publicizes the event with the relevant local media—newspapers, television stations, radio, and so on.

- **Logistics.** This job title is a broad term to cover just about everything else. Someone has to determine where the builders will arrive, where to unload the bots, where things will be staged, stored, removed, and so on.

Develop a Database

Most, if not all, robot event organizers will establish a database for potential participants, potential sponsors, vendors, and volunteers. Include the names and contacts of people and companies that may help out in other related areas—sponsoring signs, making welding and charging equipment available, catering, printing, etc. This is the business side of the activity, so remember that it is important to keep expenses at a minimum in order to break even or maybe even make a little money!

Promoting the Event

Once you have at least somewhat firm times, dates, rules, and possibly a venue, it's time to start getting builders on board. The best way is usually the most direct way. Prepare posters, mailers, e-mail, and flyers that provide general information about the event. Post your information at high schools, tech schools, science museums, college campuses—anywhere that people with a technical bent hang out. Talk to:

- High school science teachers
- High school shop teachers
- R/C hobby store owners
- Mechanical-electrical equipment surplus stores
- Science museum education staff
- Local robotics clubs

You can create a tremendous amount of interest simply by placing reasonably well designed posters and holding a few introductory seminars at local coffeehouses. Finding builders to compete in your event is generally not a problem in most areas. The desire is out there.

A prize or award should be offered to the tournament champions. Trophies are nice. Often they are handmade by the tournament organizers and take the form of giant gold-colored nuts and bolts, C-clamps, cutting blades, and so on. A cash prize to the winners of each weight class is appropriate at higher levels of competition in events where public admission is charged.

You'll find that there are lots and lots of people who are interested and have requisite technical skills but don't know how to get started. The best idea is to refer them to this book and then set up a series of informational meetings. Robot builders usually are glad to lend expertise and give advice to others, so these events can be fun and helpful for new and old hands. A reasonable registration fee will help cover initial costs and keep away the people who are not at all serious.

A Web site for your event is an important communications tool. It should provide as much information as possible. A good site will include the following pages at a minimum:

1. Main Page (introduction for your tourney with text, pictures, and links)
2. Event Information (details about your event)
3. Robot building information, including the building rules
4. Sponsors (sponsorship information)
5. Volunteer Information (information pertaining to volunteering opportunities)
6. Registration (event registration)
7. Contact Us (organization contact information)

If your Web skills are good, also include an on-line registration section.

Sweat the Details

Make sure the event you're holding will be a safe one. Remember, fighting robots are inherently dangerous if you don't take appropriate care. Ideally, the promoter will completely enclose the fighting arena in a transparent, shrapnel-proof enclosure made from polycarbonate plastic and held together firmly by a strong set of steel structural members. However, other cheaper alternatives have been tried, such as fighting in a pit with spectators looking on from a safe distance above. Other ideas include building a composite arena made from expanded metal and plastic, or simply banning spinning robots and lowering the weight limits to eliminate the more powerful and more dangerous robots. One idea for a low-cost arena utilizes an erected polycarbonate wall on just the front side and existing brick walls on the other three sides.

The promoter must implement procedures that completely eliminate radio interference between drivers and that ensure the bots are transported and operated safely. The erstwhile promoter can get a good idea of what the more established tournaments are doing by carefully studying the rules posted on tournament Web sites.

A common way of handling frequency allocation issues is to register each builder's radio control frequency in advance. Each frequency is written on a tag and each tag is hung on a large board under control of the event marshal. There can be one and only one tag per frequency. The absolute rule within the venue is that no builder may activate his or her robot without possession of the frequency tag. This precludes the possibility of two builders simultaneously operating on the same frequency and causing safety issues.

Don't forget to talk to local municipal officials to determine what permits and insurance are required to stage your event. Talk with your insurance agent about obtaining the specialized insurance necessary to stage such an event. If your agent doesn't know, he or she can probably refer you to a specialist.

Insist that each participant sign a waiver that protects the organizers from litigation should an accident occur. While no document can completely absolve the promoters of an event from liability, a signed document will help should something go terribly wrong. Remember, we're talking about mean, hard, fighting robots here.

Work Out the Tournament Event Procedures Carefully

As the event date gets closer, make sure that your staff is well prepared. Nothing ever goes exactly as planned. Some builders will not have legal or even operational bots and you won't know that until five minutes before the first match. Some of your volunteers may back out at the last minute. Have contingency plans.

Inspect and re-inspect the venue. Make certain the arena can be erected exactly as the plans describe. Arena hazards are devices that the promoter builds into the arena to provide an extra measure of danger. They can be pounding hammers, cutting blades that pop up from the floor at intervals, fire-belching pits, and trap doors. Arena hazards have a reputation for being finicky. Sometimes they work great, and sometimes not at all. It's embarrassing when your "unyielding hammer of robot death" can't even crack a walnut, but it happens. Hazards run the gamut from a motionless spike strip to giant moving carbide saws. Test everything well in advance.

Plan the pit area (the area where the builders work on their robots) carefully. Make sure everyone has access to electricity for charging batteries and plenty of clearance to move their dollied robots into and out of the pit. Make sure that each pit area is well defined with floor tape or barriers, and well marked with the name of each robot building team.

Take care of your builders; they and only they can make your event a success. Builders will appreciate the event organizer providing amenities such as soda or bottled water, welding equipment, and loaner tools. Always treat the builders with the utmost respect and do what you can to accommodate them.

Event communications are absolutely critical. Builders must know when their next match is. A large communications board listing results and upcoming matches is vital. Small events can use a bullhorn, while larger events can make use of a public address system. Some go as far as a leased two-way radio system to announce the upcoming matches, who is on deck, and so on. Be aware that builders will not wait quietly in their pit areas for you to come and get them, no matter what your rules say. They will be out and about, talking with other builders, getting coffee, fixing their bots in someone else's pit, and so on. There must be a way for you to communicate with everyone at a moment's notice to inform them of schedule changes and similar important information.

The event's technical committee should draft well thought out procedures for bringing the robots into and out of the arena. At this point, the robots (and their drivers) are charged up and ready to go, and it is quite possibly the most dangerous part of the event. Here are some suggestions for improving safety during this important time.

- Only the event marshal should be allowed to open and close the arena entry door.
- Robots must be dollied into the arena. They should not use their drive motors outside of the arena or to move into or out of the arena.
- All robots must be deactivated via their main disconnect switch or pull-out plug while being transported.
- The builder can activate the robot only when it is placed in the "robot ready" position within the arena. The builder must verify, prior to activation, that the radio is OFF.
- Radios can only be turned to ON when the event marshal verifies the arena door is shut and secure.

Make Your Event Different

There are lots of different events and ideas to try out in your tournament. Be creative and develop a buzz by developing your own format. Besides the tried-and-true one-on-one fight, consider these:

- **Melee.** Load up the arena with five, six, seven, or more robots and let them all go at it at once. Very exciting!
- **Chase the Weasel.** Buy a cheap R/C race car and let an audience member drive it around the arena while being pursued by the meanest, fastest bot in the place. If the weasel can elude the big bad bot for two minutes, the weasel wins.
- **Robot Soccer.** Set up goals at each end of the arena and let four or more robots try to push a large ball into the opposing goal.
- **Robot Agility.** Set up a course with obstacles such as sand traps, narrow bridges, low walls, and other hazards and see who can cross the course the fastest.

Smash the Bad Guy. Build a cheap robot in the form of someone evil, like Osama bin Laden, and let the other robots smash him up for a while. A great crowd pleaser!

Widen Your Scope

Investigate all avenues to make your tournament a financial success. Most of the money from the event will come from builder registration fees, spectator admissions, and, if you're lucky, event sponsorships. You can augment income by selling T-shirts, refreshments, pit passes (tickets giving attendees access to the builder pit areas), and event naming rights. Naming rights means to associate a vendor's name on all media and advertising pertaining to the event. For example, Joe's R/C Car Robot Smash.

Bookkeeping

Keep careful records. Everyone—the builders, the volunteers, the IRS—will be very interested in how and where your money came from and where it went. Of course, keep track of expenses as well as income. If keeping the books is not your forte, hire someone or get a friend to do so.

There's No Business Like Show Business

A winning robot can be a valuable commodity, and the experienced builder is advised to treat it as such. Your robot, if done well, can be a terrific opportunity for you and your sponsor. Depending on the way on it looks and operates in the arena, your robot's value may be way beyond its component value. Its image, its reputation, and its public relations and advertising buzz can be very significant assets to its owner.

Some tournament organizers insist that you sign over all rights to your robot's image to them, just for the opportunity to participate in their tournament. If the tournament is important enough to you, go ahead, but realize the following:

- The tournament's organizer, not the builder, will have control of where your robot's image can be used and who can use it. Many builders object to such arrangements because any books, Web sites, toys, stickers, magazine articles, and other avenues of publicity and profit generation are signed away, in exchange for the opportunity to participate in an event and a perhaps a shot at money from a television deal. Since it is the builder who has provided the money, expertise, time, and sweat to make the robot a champion, signing away the rights to any aspect of the robot should be done only after very careful deliberation. Obtain an advance copy of the entrance agreement and carefully look over the contract before signing.

- Surrendering control to your robot's image can have a negative impact on your ability to obtain sponsorship. If you cannot promise your sponsor free and unfettered rights to the unique visibility that your robot provides, your sponsor will go elsewhere.

- Promoters may make promises regarding the sharing of revenue from television broadcasts, toy deals, magazine royalties, and so forth. Evaluate the promises and the potential before signing away your rights. Talk to other builders and don't be afraid to ask pointed and specific questions. If the answers you receive don't make sense, then reconsider participating. Bottom line? Be very, very careful about what you sign.

Join Other Builders to Advance Your Interests

A good way for a new builder to become involved in robot building is to join an organization or club. As with any growing sport, there are several different relevant associations. Many large cities have robotics clubs where members meet to discuss all different types of robots—autonomous robots, industrial robots, sumo-style, and so on. Generally, fighting robots are not the focus in these clubs, but warrior robot builders are certainly welcome. Check out local science museums and college engineering departments to find local opportunities.

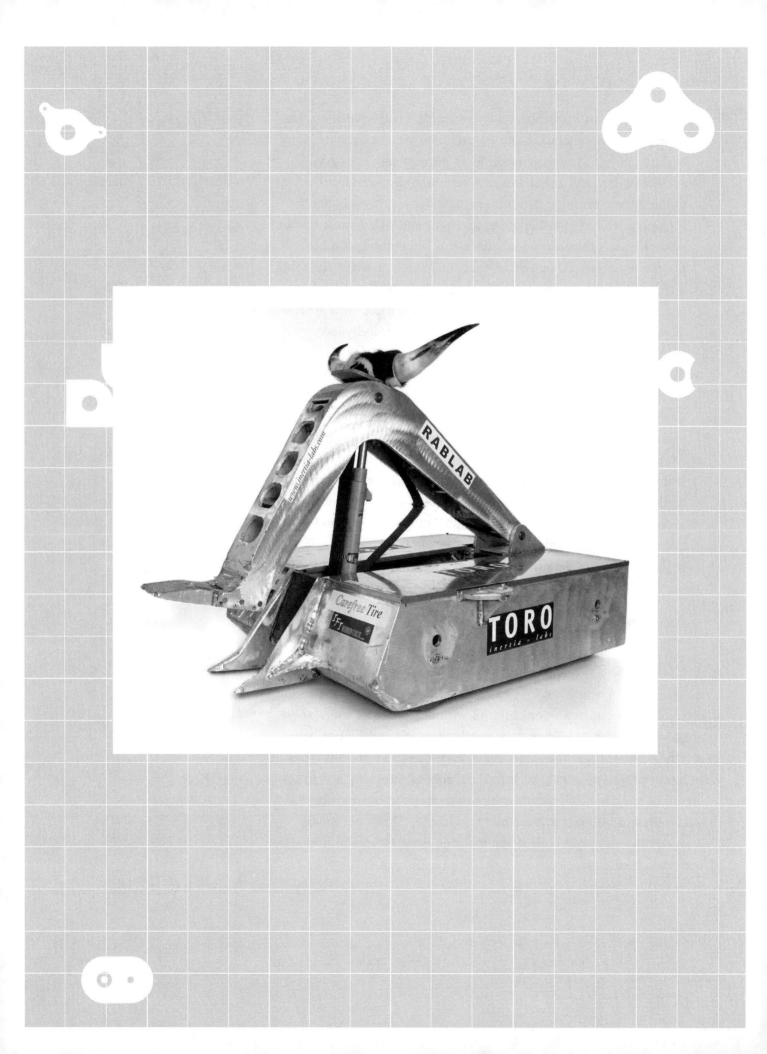

18.
GETTING ADDITIONAL INFORMATION AND HELP

The skill sets required to become a competent fighting robot builder are modest. The skill sets required to become a master fighting robot builder are enormous. Electric motor theory, applied mechanical design, control systems, basic Newtonian physics, and materials science are just some of the things a good robot builder knows more than a little about. Certain people, mostly fictional, just seem to innately know a lot of this stuff. But most people need help in at least some areas.

You've made the initial investment in learning about robot building by buying this book. However, as your robot building skills grow, and your ideas become more involved, it is important to cast a wide net on a variety of topics. The scope quickly becomes so wide that no single book, however comprehensive, can provide everything.

At some point, you'll have questions or problems that can't be easily solved by the information found here. This chapter is a guide to obtaining the additional information you need.

Experienced robot builders have a reputation for being helpful to other builders. This is basically true, but with important qualifications. The builders and designers of the most successful robots—the ones you see on television—receive vast numbers of e-mail messages from

Toro COURTESY BATTLEBOTS, INC., AND INERTIA LABS

wannabe robot builders, most of them containing vague or unstructured questions. The experienced builders, nice as they are, generally don't have the time to respond to individuals asking the following types of questions:

1. Poorly formulated and/or nonspecific questions such as "What's the best type of weapon for my vapor bot?" or "How big a motor do I need in order to defeat XYZ robot?"
2. Requests for sponsorship or money. You may have a great idea for a robot, but no potential sponsor is going to care until you can back up your ideas with results on the battlefield.
3. Questions that could be answered by the questioner with a minimum to moderate amount of research on the Internet.
4. Questions about what is the best robot, who is the best builder, which is the best competition, and so on. Most good builders are simply too busy with their own projects to have enough time to answer questions like these.

So where can you go for additional information? Thankfully, there are many resources available to the intrepid builder.

Robot Parts Suppliers

Many, but not all, vendors of robotic parts (DC motor suppliers, electronic speed controllers, NiCad or SLA battery suppliers) will be happy to provide specific information on the parts they sell, via the Web or e-mail. They are willing to give information, but only up to a point. If they perceive by your phone or e-mail questions that you are completely clueless, they'll clam up. They don't want to waste too much time trying to educate someone who isn't capable of actually buying their parts and building something.

General Materials Vendors
Getting good information from non-robot-oriented material vendors can be a hit-or-miss affair. If you go into a recycled auto

parts yard and ask about using a starter motor in a Ford pickup truck, you can get pretty good information. But if you ask the same person about how to rewire the starter motor so it will be reversible, well, that's really expecting too much. Similarly, the person at the metal supply store may be quite familiar with the machining properties of mild steel and normal aluminum alloys. But does this person know much about titanium or aircraft aluminum alloys? Maybe or maybe not.

Hobby Stores

An exception is hobby stores that sell model airplanes and cars. If you visit them on a day that's not too busy, they often provide reasonably good insight into the R/C world. Keep in mind, however, that they are focused on R/C vehicles, not robots, so their knowledge of anything beyond standard R/C may be very limited. Still, it is a good place to start.

At minimum you'll have an opportunity to become familiar with AM, FM, and FM-PCM radios; DC airplane motors; and NiCad batteries. You may not choose to use any of these parts in your own robot, but a background in basic R/C is valuable.

Builder Web Sites

The Internet is a very good place to obtain all sorts of information on the sport. Appendix I in this book lists several Web sites with reasonably reliable information on many topics of interest to fighting robot builders. Also, visit the FAQ pages at www.building-bots.com.

Internet Forums

Another good spot for detailed information on a huge number of fighting robot-related subjects are the Delphi forums at www.delphiforums.com. There are a number of relevant forums, including the huge Battlebot builders forum, the Robotica forum, the Robot Wars discussions, and several others. Most forums contain discussion threads on motor selection, radio control, speed control, machining, and other robot-building topics.

Forums can be very difficult to use, however. Remember that there are many, many different threads or discussions taking place, and it is difficult to sift through 50,000+ different messages to find the information you actually care about. The forum host provides search engines that allow you to search by phrase, by forum, by time period, and by poster name. These tools do help, but you still need to have a pretty good idea of what you're searching for to be successful. When you do find something that appears relevant, it may or may not be accurate, so you always have to determine the veracity of each nugget of information.

Forums are known as being lively places of information exchange. There also tends to be a lot of whining and complaining. People who follow the various forums closely may get testy about postings requesting information that has been provided in the past. Sometimes the exchanges escalate into name calling and insults. In order to avoid being called clueless or juvenile by the classless types who love insulting—or "flaming"—the inexperienced, read the forum for a while without posting your own messages. You will get an idea of the way the forum operates. Check the past postings and FAQs to see if a particular question has already been asked and answered.

BATTLEBOTS 2001

Treasure Island, San Francisco, California

There is a big difference between Battlebots and everything else in the world of fighting robots. Automobile racing has the Indy 500, horseracing has the Kentucky Derby, and combat robotics has Battlebots. In terms of the number of competitors, the quality of the robots, and the size of the audience, Battlebots is certainly the largest and probably the most important tournament.

It has been held in various venues, most recently at Treasure Island, in the middle of San Francisco Bay. This manmade island was once used as the city's airport but was eventually taken over by the navy. In the island's huge Hangar Number 3, the Battlebots' television production company, which produces the show for a cable television network, has set up the competition.

During the fall 2001 Battlebots tournament, security was tight as rumors flew about the Bay Bridge being a possible target for terrorists. But the event was held anyway, and no one inside seemed to care much about the vague threats. "Let 'em come and try something," one builder said, "my robot will kick their ass."

Television cameras, lights, and aluminum bleachers surround the arena area, where the Battlebox is located. The rest of the building is the pit area, where a hundred tables are set up for builders to make changes and repairs to their robots. Unlike the crude signage and naked incandescent bulbs at Bot Bash and Mechwars, Battlebots is very much a Hollywood-style event. A three-story-high, 80-foot-long banner hangs from the west wall and proclaims "BATTLEBOTS" in huge, stylized letters. The producers thoughtfully provide bottled water, doughnuts, fruit, and other snacks for the builders who, unused to the concept of free food, gobble it all down as if they can't believe their luck.

The actual Battlebots tournament goes on for a full week. The first few days are for check-in and safety inspection. Each robot is weighed on a digital scale. The best builders bring their bots in at *exactly* the maximum weight for any weight class. An unfortunate few weigh in a pound or two over and, to make their bots legal, are forced to perform an emergency metal-ectomy. The ensuing process is unpleasant for builder and robot alike, often hurriedly conducted like a Civil War battlefield amputation. These weight-reducing surgeries invariably result in what appears to outsiders as unexplainable holes and voids in the robot's body armor.

After weigh-in, the bots go to a special safety-check area where Battlebots staff check each bot for rigorous adherence to the safety rules. The rumor is that this is most dangerous place on the planet, as untested robots can go ape during this initial run-up, resulting in a high-speed spray of metal detritus throughout the test area. After inspection, the safety-checked bot returns to its pit area to await the scheduling and seeding process for its first trip to the Battlebox.

The Battlebox is a 16-foot-high polycarbonate plastic enclosure topped by a metal superstructure to hold the expected metal carnage within. The Battlebox's footprint is about 50 feet long by 50 feet wide. Within are the arena's hazards, the most noticeable of which are four remote-controlled hammers located at each corner, which are there to pound the daylights out of anything unfortunate enough to wander beneath.

There are two types of matches: qualification rounds and TV rounds. The qualification rounds are first. They are two minutes long and take place in the Battlebox, but no hazards, save the row of steel spikes that line the arena perimeter, are present. New builders and veteran builders with new robots all must go through the qualification rounds. (Ranked robots, those with a successful track record, skip the qualification rounds and go directly to the TV rounds.) Those who qualify by winning the early rounds advance to the TV rounds, where the real fun begins.

The TV rounds are three minutes long and take place on the last two days of the tournament. All arena hazards are working, and the television celebrities signed on to work the event are present.

More than any other sport, there is a *huge* difference between winning and losing. Battlebots is a single-elimination tournament—one loss and you're toast. Most builders don't really mind losing in a late match; after all, there is only one winner out of hundreds of entrants. But losing in an early round is horrible, made even more horrible if the loss was a fluke or just bad luck.

One builder told me he spent $10,000 on his middleweight robot. He packed the thing with incredibly expensive, incredibly powerful DC motors, armored it with a titanium plate, and installed a 900 MHz computerized radio system and motor controller so precise it could pick up a sewing needle. He lasted only 35 seconds into his first match against something that looked like a Weber grill on roller skates because the six-dollar emergency cutoff switch he installed decided to turn itself off and shut down all power. As I said, it's one loss and you're done, and this guy was sick with disappointment. He told me this story as I sat next to him in the stands, only minutes before he was thrown out of the venue for sneaking an unauthorized photograph, with his disposable camera, of himself with his arm around Carmen Electra.

As long as you win, you're the center of attention. No matter how awkward you are on camera, no matter how tongue-tied you get during your interviews with second-tier celebrities such as Carmen Electra or comedian Bill Dyer, you are the center of attention. A friend of mine, a completely unknown builder from a small town in central Wisconsin, came out of nowhere to win match after match in the super-heavyweight division. Obviously uncomfortable in the limelight, he more than held his own in the arena, clobbering everybody until losing on points in the final championship match to a very well known and absurdly powerful lifting robot. The producers did what they could to coax charm and personality out this guy during the interviews. But, like many others in the sport, he has the star appeal of a computer programmer and the wit of an engineer, neither of which scream out to television producers "here is a good interview."

But the producers know what they're doing. They placed the guy's grinning three-year-old kid on his daddy's shoulders and let the kid's smile do the talking for the camera. Even a kid knows it's lots better to win.

APPENDIX I: PART SUPPLIERS

General Parts Suppliers

National Power Chair
4851 Shoreline Drive
P.O. Box 118
Mound, MN 55364
(800) 444-3528
www.npcinc.com
DC electric motors, wheels, hubs, and electronics.

MECI
340 E. First Street
Dayton, OH 45402
(800) 344-4465 or (937) 461-3525
www.meci.com
Varied, changing selection of surplus electronic parts, motors, miscellaneous mechanical, and drivetrain parts.

McMaster-Carr
600 County Line Road
Elmhurst, IL 60126
(630) 933-0300
Sales@mcmaster.com
www.mcmaster.com
Raw materials, fasteners, drivetrain parts, tools, fluid power equipment, motors, and electrical parts.

W.W. Grainger, Inc.
100 Grainger Parkway
Lake Forest, IL 60045-5201
(888) 361-8649
www.grainger.com
Fasteners, drivetrain parts, tools, fluid power
equipment, gas engines, motors.

Technobots Ltd
P.O. Box 227
Totton, England SO40 3WE
44 (0)23 8086 3630
www.technobots.co.uk
Motors, radio parts, speed controllers, and drivetrain
parts.

Team Delta
www.teamdelta.com
Electronic speed controllers, radio-controlled relays,
motors, main cutoff switches, many specialized parts.

American Science & Surplus
3605 Howard Street
Skokie, IL 60076
(847) 982-0870
www.sciplus.com
Varied selection of electronic parts, motors, wheels,
and mechanical parts.

C & H Sales
2176 E. Colorado Boulevard
Pasadena, CA 91107
(800) 325-9465 or (626) 796-2628
aaaim.com/CandH
Varied selection of electronic parts, motors, wheels,
and mechanical parts.

Small Parts, Inc.
13980 N.W. 58th Court
P.O. Box 4650
Miami Lakes, FL 33014-0650
(800) 220-4242
www.smallparts.com
Fasteners, drivetrain parts, and raw materials.

Team Whyachi LLC
P.O. Box 109
Dorchester, WI 54425
(888) 937-8271 or (715) 654-5000
www.teamwhyachi.com
DC motors, drivetrain parts, and wheels.

Batteries

Battery Stuff
P.O. Box 1703
Rogue River, OR 97537
(800) 362-5397
www.batterystuff.com
Batteries and chargers.

Ballistic Batteries
P.O. Box 30344
Phoenix, AZ 85046-0344
(602) 493-3717
www.ballisticbatteries.com
NiCad batteries.

Hawker Batteries
www.hepi.com
617 N. Ridgeview Drive
Warrensburg, MO 64093-9301
(660) 429-2165
Sealed lead acid batteries.

Battlepacks
305 9th Street
Carrollton, KY 41008
(502) 639-0319
NiCad batteries.

Computer Aided Design (CAD) Services

Caztek Inc.
5336 Elliot Ave. S
Minneapolis, MN 55417
www.caztek.com
(612) 825-1272
CAD and precision machining services for the robot builder.

Mike Konshak's Robot Dojo
www.robotdojo.com
Large selection of CAD drawings of many commonly used parts. Currently, downloads of drawings are free. Good resource!

Drivetrains

W. M. Berg Co.
499 Ocean Avenue
East Rockaway, NY 11518
(516) 599-5010
www.wmberg.com
Mechanical and drivetrain parts.

KW Marketing
1488 Mentor Avenue
Painesville, OH 44077
(440) 357-5569
www.kwmarketing.com
Wheels, tires, and trailers.

Northern Tool & Equipment
www.northerntool.com
P.O. Box 1219
Burnsville, MN 55337-0219
(800) 221-0516
Wheels, hubs, drivetrain parts, and gasoline engines.

Boston Gear
14 Hayward Street
Quincy, MA 02171
(617) 328-3300
www.bostgear.com
Mechanical and drivetrain parts.

Servo-Link, Inc.
5356 W. Vickery
Fort Worth, TX 76107
(817) 732-GEAR
www.servolink.com
Plastic gears, sprockets, and drivechains.

Allied Devices, Inc.
325 Duffy Street
Hicksville, NY 11801
(516) 935-1300
www.allieddevices.com
Mechanical and drivetrain parts.

Electronic Speed Controllers

Vantec
460 Casa Real Plaza
Nipomo, CA 93444
(888) 929-5055 or (805) 929-5056
www.vantec.com
Electronic speed controllers for DC motors.

Robot Power
31808 8th Avenue South
Roy, WA 98580
(253) 843-2504
www.robot-power.com
Open Source Motor controller kits.

Diverse Electronics Services
1202 Gemini Street
Nanticoke, PA 18634-3306
(570) 735-5053
members.tripod.com/~divelec/
Electronic speed controllers.

Novak Electronics, Inc.
18910 Teller Avenue
Irvine, CA 92612
(949) 833-8873
www.teamnovak.com
Small electronic speed controllers.

IFI Robotics
9701 Wesley Street, Suite 203
Greenville, TX 75402
(903) 454-1978
www.ifirobotics.com
Remote control radio systems, electronic speed controllers.

Fluid Power Supplies

Palmer-Pursuit
3951 Development Drive, #3
Sacramento, CA 95838
(916) 923-9676
www.palmer-pursuit.com
High-pressure air and CO_2 tanks and equipment.

The Paintball Store.com
6513 Storage Drive
Amarillo, TX 79110
(806) 467-0277
www.thepaintballstore.com
HPA and CO_2 tanks and equipment.

Fabco-Air, Inc.
3716 N.E. 49th Avenue
Gainesville, FL 32609-1699
(352) 373-3578
www.fabco-air.com
Industrial pneumatic tanks and cylinders.

Motors

MicroMo Electronics
14881 Evergreen Avenue
Clearwater, FL 33762-3008
(800) 819-9516 or (727) 572-0131
www.micromo.com
Small DC motors.

Astro Flight Inc.
13311 Beach Avenue
Marina Del Rey, CA 90292
(310) 821-6242
E-mail: info@astroflight.com
www.astroflight.com
Brushless DC motors, motor controllers, and NiCad batteries.

4QD
30 Reach Road
Burwell, Cambridgeshire CB5 0AH
United Kingdom
www.4QD.co.uk
Motors and motor controllers.

Aveox Inc.
31324 Via Colinas, Suite 103
Westlake Village, CA 91362
(818) 597-8915
www.aveox.com
Brushless DC motors and motor controllers.

Magmotor
121 Higgins Street
Worcester, MA 01606
(508) 856-7220
www.magmotor.com
DC motors.

CMACM Technologies
214 Ruby Street
Sheridan, MT 59749
(406) 842-5339
www.3rivers.net/~cmac/cmac1.htm
DC motors.

Electrol Co.
P.O. Box 29
York, PA 17405-0029
(717) 848-1722
www.electrolco
DC motors and motor controllers.

Poly-Scientific
213 N. Main Street
Blacksburg, VA 24060
(800) 336-2112
www.polysci.com
Brush and brushless DC motors and motor controllers.

Pittman Motors
343 Godshall Drive
Harleysville, PA 19438
(877) 748-8626 or (215) 256-6601
www.pittmannet.com
DC Motors.

Leeson Electric Corporation
www.leeson.com
P.O. Box 241
Grafton, WI 53024
(262) 377-8810
Large DC electric motors.

EV Parts, Inc.
(888) 387-2787 or (425) 672-7977
www.evparts.com
Larger DC electric motors and controllers.

Harbor Freight Company
(800) 423-2567
www.harborfreight.com
Low-cost electric drill motors.

Radio Control Equipment

Tower Hobbies
P.O. Box 9078
Champaign, IL 61826-9078
(800) 637-6050 or (217) 398-3636
www.towerhobbies.com
Radio-controlled transmitters, receivers, and servomotors.

Horizon Hobby
4105 Fieldstone Road
Champaign, IL 61822
(217) 352-1913
www.horizonhobby.com
Radio-controlled transmitters, receivers, and servo motors.

Raw Materials

Online Metals
366 West Nickerson, Lower Level
Seattle, WA 98119
(800) 704-2157
www.onlinemetals.com
Metal and plastics.

Specialized Parts

Robotparts
37 W. Wheelock Street, Suite D
Hanover, NH 03755
www.robotparts.org
Custom machined metal parts.

West Marine
P.O. Box 50070
Watsonville, CA 95077-0070
(800) 262-8464 or (831) 761-4800
Heavy-duty electrical switches for main cutoff switches.

Cruel Robots
32547 Shawn Drive
Warren, MI 48088
www.cruelrobots.com
Wheels, axles, reducers, hubs, and custom machined parts.

Tools

Sherline Products Inc.
3235 Executive Ridge
Vista, CA 92083-8527
(800) 541-0735 or (760) 727-5857
www.sherline.com
Small machine tools, including mills and lathes.

APPENDIX II: RADIO FREQUENCY CHART

27 MHz	
Channel	Frequency
A1	26.995
A2	27.045
A3	27.095
A4	27.145
A5	27.195
A6	27.255

50 MHz	
Channel	Frequency
00	50.800
01	50.820
02	50.840
03	50.860
04	50.880
05	50.900
06	50.920
07	50.940
08	50.960
09	50.980

72 MHz

Channel	Frequency	Channel	Frequency	Channel	Frequency	Channel	Frequency
11	72.010	24	72.270	37	72.530	50	72.790
12	72.030	25	72.290	38	72.550	51	72.810
13	72.050	26	72.310	39	72.570	52	72.830
14	72.070	27	72.330	40	72.590	53	72.850
15	72.090	28	72.350	41	72.610	54	72.870
16	72.110	29	72.370	42	72.630	55	72.890
17	72.130	30	72.390	43	72.650	56	72.910
18	72.150	31	72.410	44	72.670	57	72.930
19	72.170	32	72.430	45	72.690	58	72.950
20	72.190	33	72.450	46	72.710	59	72.970
21	72.210	34	72.470	47	72.730	60	72.990
22	72.230	35	72.490	48	72.750		
23	72.250	36	72.510	49	72.770		

75 MHz

Channel	Frequency	Channel	Frequency	Channel	Frequency	Channel	Frequency
61	75.410	69	75.570	77	75.730	85	75.890
62	75.430	70	75.590	78	75.750	86	75.910
63	75.450	71	75.610	79	75.770	87	75.930
64	75.470	72	75.630	80	75.790	88	75.950
65	75.490	73	75.650	81	75.810	89	75.970
66	75.510	74	75.670	82	75.830	90	75.990
67	75.530	75	75.690	83	75.850		
68	75.550	76	75.710	84	75.870		

APPENDIX III RULES FOR A TYPICAL ROBOT TOURNA- MENT

I. Robot Construction

a. Radio Systems. Teams must register the frequency of their controller with the Event Marshal. There are specific channels for controlling flying devices. Use of radios not licensed for terrestrial communication is against FCC regulations.

b. Compressed Gas and Hydraulic Systems. Supply tank pressures are not to exceed 1,000 psi. Higher pressures may be allowed if a waiver is granted. The Event Organizer or Event Marshal can grant waivers after physical inspection of the system. Low-side pressures, i.e., downstream of the main tank regulator, may not exceed 150 psi. Again, higher pressures may be allowed if the Event Organizer or Event Marshal grants a waiver after physical inspection of the system.

c. Compressed Gas and Hydraulic Systems, Continued. The builder must be able to depressurize the entire system, high-pressure side and low-pressure side, through a purge valve on either side. Pressure gauges must be present on both high side and low side. All pressurized components must be manufactured for the purpose of holding or controlling pressurized fluid and must be rated as such by a U.S. Government rating agency. Tank heaters are not allowed on CO_2 systems. All pneumatic and hydraulic fluids must be nonflammable.

d. Petroleum Fuels. Machines may carry chemical fuel for no more than 4 minutes of continuous operation or 3 fluid ounces, whichever is less. Non-gasoline fuels must be pre-cleared with the judges. Fuel tanks must be sturdy enough to withstand any possible impact.

e. Batteries. Only sealed batteries (SLA gel cells and NiCads) are acceptable.

f. Cutoff Switch or Immobilization Plug. Every robot must have an easily actuated cutoff switch or plug that can be operated quickly.

g. Emergency Power Cutoff

    i. Builders must be able to render their robot harmless in less than 7 seconds.

    ii. A robot is only considered safe when its drive and weapon systems are completely unpowered and not moving.

II. Robot Weight Classes

a. Ultralightweight: Wheeled robots may weigh up to 15 pounds.

b. Lightweight: Wheeled robots may weigh between 15 and 45 pounds.

c. Middleweight: Wheeled robots may weigh up to 120 pounds.

d. Walking and Stomp Bots: Robots that travel by means of linear actuated legs instead of wheels; such robots may exceed the above weight limits provided they meet the following criteria:

    i. May weigh up to 20 percent more than other robots in their class.

    ii. Cam-actuated legs may or may not be allowed in the weight advantage. Cam-actuated robots should be inspected prior to the event by the Event Marshal for conformation determination.

III. Arena Egress and Ingress

a. Robots must fit through the arena door. The door opening is 36 inches wide by 86 inches high.

b. Robots must be carried or dollied through the door of the arena.

c. Robots must be fully assembled prior to entering the arena, except for the connecting the main cutoff switch connector, if so equipped.

d. Participants are to enter and exit the battle cage quickly.

e. The builder is responsible for having enough people/equipment to move the robot into and out of the arena area. Robots cannot be driven into or out of the arena.

f. Start Up. Drivers may turn on only when given the signal to power on by the Event Marshal.

g. Power Off. Drivers must immediately power down when instructed to do so by the Event Marshal.

IV. Robot Mobility

a. All machines must be able to travel a distance of 35 feet at a minimum speed of 0.5 miles per hour.

b. Robots must be able to make it up a 5 percent grade and be capable of traversing the small cracks and ridges in the arena floor.

c. No flying robots are allowed.

V. Weaponry

a. All cutting edge or sharp weapons, such as spikes and saw blades, must have secure covers for transport. Covers must be secured in place by tape or elastic at all times except within the arena.

b. No intentional radio interference is permitted.

c. No sensory interference with other drivers (e.g., sirens, strobe lights) is permitted.

d. No explosives are permitted.

e. Use of liquids or chaff as weapons is not allowed.

f. Weapons must not interfere with an opponent's radio signal. For example, spark generators that cause radio interference for any significant interval are not allowed.

g. Net or rope weapons must be 5 feet long or less and cover a maximum area of 4 square feet or less.

h. Moving saw blades, hammers, and other mechanisms must have safety devices that restrain their motion between matches.

i. No projectile weapons are permitted.

VI. Disqualification

a. Failure to adhere to these rules and regulations may result in disqualification.

b. Any machine that is considered by the Event Marshal, at his or her sole discretion, to be dangerous to spectators or opponents will be disqualified.

VII. Team

a. Size

i. Teams will consist of six people or fewer and must have an adult member.

ii. Four or fewer members can be designated "crew" and allowed into the pits.

b. Liability Disclaimer. All opponents and spectators assume the risk of participation in the event. The organizers of the event disclaim all warranties and liability, express or implied, regarding claims or damage resulting from the event.

c. Communication During the Event. Teams must have one of the following means of being immediately contacted by the Event Marshal staff during the event:

i. A cell phone.

ii. A pager.

d. Team Behavior

i. Any behavior that shows disregard for the safety of others will result in ejection from the tournament.

ii. Judges' decisions are final. Arguing and poor sportsmanship are not allowed and are grounds for disqualification and ejection.

VIII. Event

a. In-arena failures prior to match start. Robots that do not operate at the start of the match will be declared the loser.

b. Rescheduling. In the spirit of sportsmanship, the match may be rescheduled to a later time at the sole discretion of the opponent. If granted, the match will simply be rescheduled to a mutually agreeable later time slot.

IX. Judging

a. Points will be awarded for damage to the opponent, aggressiveness as determined by the judges, driving skill, and audience reaction.

b. Score sheets will be used to note scoring events, in this order of importance:

    i. Match ending event. Damage is sustained such that the robot's driver cannot make the robot move through radio control.

    ii. Significant damage to exterior.

    iii. Impaired drive motion.

    iv. Impaired weapon.

    v. Aggressiveness/Driving Skill/Appearance.

c. Immobilization of the opponent without damage (i.e., flipping or pinning) will not result in an automatic loss. When such an event occurs, the mobile robot will have 15 seconds to attack before the match is temporarily halted so the immobilized bot can be righted or unpinned and the match restarted. The clock stops while the arena door is open.

d. At the end of the match, if no damage is registered in the judge's evaluation, or if damage assessments are equal, the judges will award the match based on aggressiveness, driving skill, or other criteria at the judges' discretion.

X. Awards

a. The winner of each weight class will receive an award.

b. The winner of each weight class will receive a cash award in the amount of $100.

c. The second-place winner will receive a cash award in the amount of $75.

d. The third-place winner will receive a cash award in the amount of $50.

XI. Special Events

a. Time and competitor interest permitting, there may also be general melees, weasel races, and robot soccer matches.

GLOSSARY

ALLOY Two or more metals or elements combined to form a new metal with different material properties.

ANTWEIGHT The smallest competitive robot class, generally weighing less than two pounds.

BEARING Any part that bears, or supports, another (usually a rotating) part; a bearing is commonly thought of as a component that uses balls in a metal raceway to reduce rotating friction.

BEVEL A slanted surface machined onto another surface.

BMW A type of robot that only uses a ram for striking opponents; abbreviation for "batteries, motors, and wheels."

BOSS The raised surface of a circular outline.

BRASS An alloy of copper and zinc, or copper with zinc and lead.

BRAZE To join two pieces of metal by using a hard solder like brass or zinc; brazing provides a stronger mechanical joint than soldering but not as strong than a welded one.

BRONZE An alloy of copper and tin.

BURR A rough or jagged edge caused by the cutting, shearing, or other fabrication process; good machinists will usually remove burrs before handling a part.

BUSHING A hollow, cylindrical sleeve used as a bearing or guide.

CAM A carefully shaped machine part mounted on a revolving shaft to change rotary motion to linear motion.

CASTING A part made by pouring molten metal into a mold.

CHAMELEON BOT A robot that can use different types of bolt-on weaponry, depending upon the characteristics of its specific opponent.

CHAMFER An edge that has a bevel.

CHANNEL (1) The frequency number used by the transmitter to send signals to the radio receiver; (2) The number of functions a given R/C system can control; for example, an 8-channel radio has 8 available slots that can used for 8 separate robot controls, including motors, weapons, and accessories.

COUNTERBORE The enlarged end of a hole.

COUNTERSINK A conical enlargement of the end of a hole.

COUPLING The mechanical attachment of a load to a motor.

DENSITY The mass of substance per unit volume.

DIE A hard tool that cuts external screw threads.

DOWEL A cylindrically shaped pin used for fastening parts together.

DRAFT The tapered sides on a part that allow it to be easily removed from a mold.

FILLET The rounded corner between two surfaces.

FIT The tightness or looseness between meeting parts (interference or slip).

FIXTURE A holder for a work piece.

FLANGE A metal lip or extension from a surface.

FLIPPER-BOT A robot equipped with an electric, pneumatic, or hydraulic lifting arm that is built to flip over opposing robots; sometimes called a lifter.

FREQUENCY ALLOCATION The Federal Communications Commission (FCC) has allocated the 72 MHz band for use by R/C flying machines; this band is divided up into many different channels in which you can operate a radio system (see Appendix II); the FCC has allocated the 75 MHz frequency range (75 410 through 75 990) for ground model use only —robots as well as R/C cars; there is also a 50 MHz band, but only licensed HAM radio operators may use it.

FUBAR A robot design or prototype that fails miserably; acronym for "fouled up beyond all recognition."

GAUGE A device for determining whether a specified dimension on an object is within a specified tolerance.

GEAR RATIO The ratio of one gear diameter to the mating gear diameter; this ratio determines the input and output speeds of a geartrain.

GLITCHING If multiple radios transmit on the same frequency, or channel, glitching occurs in the active receiver on that channel; this is due to conflicting signals sent by the two radios.

GOATSCREW A very chaotic or unfair state of affairs, as in "That whole tournament was very disorganized. What a goatscrew."

JIG A guide for a cutting tool or a fixture.

KEY A small metal bar, placed in a groove in a shaft, that allows the shaft to transmit torque to another part, such as a wheel or gear.

KEYWAY A groove or slot in a shaft or hub into which a key is placed.

KNURL A series of shallow cuts that roughen a cylindrical surface so it can be more easily turned by hand.

MILL To machine a part on a milling machine using a rotating toothed cutter and movable table.

MITER A 45-degree bevel.

MIXING Combining electronic signals to allow a single joystick to control the operation of two or more radio output channels; mixing simplifies robot driving and allows more complex maneuvers.

MODULUS OF ELASTICITY A measure of a material's stiffness; also known as Young's Modulus.

NORMAL SHOCK A rapid, jarring change of direction in the direction opposite to the original motion; what a robot feels when it rams straight on into something.

PLASTIC DEFORMATION A permanent set or deflection in a part.

PLATE (1) To electrochemically coat a metal object with another metal; (2) Sheet metal with a thickness of ¼ inch or more.

POP An acronym for "pile of parts"; the intermediate stage of robot building between the time you sketch out the idea and the time it is actually finished; for example, "Is your robot ready?" "No, it's still a POP."

POWER The rate of doing work as a function of time; the more work a robot does for a given unit of time, the more power it consumes.

PRECIPITATION HARDENING An aluminum heat-treating process; over time, a chemical process occurs in precipitation hardening aluminum alloys that changes the metallic structure and makes the aluminum stronger and tougher.

PUNCH To pierce thin material by pressing a tool through it.

REAM To fabricate a hole to a very accurate diameter.

SAFETY FACTOR The ratio of a material's handbook mechanical limits to the highest calculated stresses or strains it will endure during actual use.

SERVO OR SERVO MOTOR An electromechanical device that moves the mechanical controls of a robot according to commands from the radio system.

SHEAR STRESS The highest torsional (twisting) stress a material can tolerate.

SHEAR MODULUS The measure of a material's resistance to deformation from shear.

SHIM A thin plate of material used between two surfaces to adjust the distance between them; shims are often used to fix alignment problems between sets of pulleys or gears.

SNAFU An unsuccessful robot design or trial; U.S. Army acronym for "situation normal: all fouled up."

SOLUTION HEAT TREATMENT An aluminum heat-treating process involving quenching the heated part in water or oil.

SPINNER A type of high kinetic energy robot weapon wherein the robot's external shell rotates at high speed, causing damage to whatever it touches.

SPLINE A long keyway with several torque-transmission surfaces.

STRAIN The dimensional changes resulting from stress.

STRESS A unit of force divided by the area over which it is applied.

T-LAR An inaccurate, but frequently used, measuring procedure that uses eyeballing and guessing; acronym for "that looks about right."

TAP To cut bolt threads in a hole.

TEMPER To reduce the brittleness in a hardened metal part by heating it.

TENSILE STRESS The highest stress a material can tolerate before it pulls apart.

THWACK-BOT A two-wheel drive robot that uses a spinning, heavy appendage for a weapon.

TOLERANCE An allowance for variation from a given dimension in a part.

TOUGHNESS The capability of a metal to absorb highly localized and rapidly applied stresses.

ULTIMATE STRESS The highest stress before failure occurs.

VAPOR-BOT A robot that only exists on paper or in someone's imagination.

WASHER A ring of metal used to form a seat for a bolt and nut.

WEDGE-BOT A robot that is designed to crash into other robots using its wedge-shaped front.

WELD To join pieces of metal by heat or pressure.

WORK Force acting through a distance.

YIELD STRESS The highest stress before plastic (nonrecoverable) deformation occurs.

YOUNG'S MODULUS The ratio of stress to strain in the elastic range of a material.

INDEX